A GARDENER TOUCHED WITH GENIUS

The Life of Luther Burbank

REVISED EDITION

Also by Peter Dreyer

THE FUTURE OF TREASON

A BEAST IN VIEW

MARTYRS AND FANATICS

A GARDENER TOUCHED WITH GENIUS

The Life of Luther Burbank

Peter Dreyer

REVISED EDITION

UNIVERSITY OF CALIFORNIA PRESS
Berkeley • Los Angeles • London

University of California Press
Berkeley and Los Angeles, California

University of California Press, Ltd.
London, England

Library of Congress Cataloging in Publication Data

Dreyer, Peter, 1939–
A gardener touched with genius.

Bibliography: p.
Includes index.
1. Burbank, Luther, 1849–1926. 2. Plant breeders—
United States—Biography. I. Title.
SB63.B9D73 1985 635′.092′4 [B] 84-24122
ISBN 0-520-05116-5

Printed in the United States of America

1 2 3 4 5 6 7 8 9

In memory of
Walter Lafayette Howard (1872–1949)
and George Harrison Shull (1874–1954),
who in large part made this book possible

CONTENTS

My measure for all subjects of science as of events is their impression on the soul. That mind is best which is most impressionable. . . . But sensibility does not exhaust our idea of it. That is only half. Genius is not a lazy angel contemplating itself and things. It is insatiable for expression. Thought must take the stupendous step of passing into realization.

—Ralph Waldo Emerson,
"Natural History of Intellect"

PREFACE

> I, on my side, require of every writer, first or last, a simple account of his own life, and not merely what he has heard of other men's lives; some such account as he would send to his kindred from a distant land; for if he has lived sincerely it must have been in a distant land to me.
>
> —Henry David Thoreau, *Walden*

"That is good," Luther Burbank noted in his autobiography. "I will write of this Harvest of the Years of mine to you, my kindred, as though from a distant land!" It was of the utmost importance to him to have lived sincerely. He sought unremittingly to prove, not so much to others as to himself, that he had done so. In the end, he had probably not quite succeeded, and Thoreau's irony (to which he was very likely blind) is no more than appropriate.

Despite the façade of simplicity that deceived critics and customers alike, he was an immensely intricate character. It was his great misfortune that he was compelled to pursue the seedsman's trade to support the experimental work to which he by choice devoted his life. It was his misfortune, too, that he possessed a mysterious charisma that drew sycophants and eulogists like bees to nectar, and was never able to resist playing to their adulation.

In his own lifetime, millions of words were written about him, hardly any of them critical. To this day there has been only one book published that assesses him in terms of his real merits and real faults, Walter L. Howard's monograph "Luther Burbank: A Victim of Hero Worship." Appearing as a double number of the periodical *Chronica Botanica* in the winter of 1945, it never reached the general public and is now hard to come by. It was written, says Howard's son, Walter E. Howard, "to keep Burbank from becoming a victim of hero worship, with the scientific community discrediting his contributions."[1] Its author, then emeritus professor of pomology and late director of the College of Agriculture of the University of California at Davis, had spent ten years privately researching Burbank's plant contributions (these results were separately published as Bulletin Number 691 of the University of California Agricultural Experiment Station at Berkeley) and had come to realize the degree to which his genuine achievements had been eclipsed by his overinflated popular reputation. The legend obscured the man. Horticultural scientists, the only ones in a position to judge his work, came to dismiss him as no more than a commercial breeder with a talent for publicity and as something of a fraud into the bargain. As the geneticist Donald Jones observed, reviewing Howard's study:

> Any writer about Burbank and his work has the difficult task of sifting the exaggerated claims and misstatements arising both from Burbank's uncritical estimate of his work and the publications of writers who knew little or nothing about the scientific aspects of the subject. On the other hand, there is the overly critical attitude of specialists in the seed and nursery trade and professional geneticists and biologists, who properly resent and distrust the nonsense about Burbank that has been too prevalent in the popular press. The background and setting of the conflicting interests that swirled about Burbank during the latter part of his life make the real accomplishments of a long life of hard work begin to stand out.[2]

Now curiously enough, Jones (for many years head of the Department of Genetics at the Connecticut Agricultural Experiment Station) had himself written a book about Burbank during the latter's lifetime. Although giving Burbank his due—uniquely so in acknowledging his priority in reporting true-breeding species hybrids—it seemed excessively critical to lay readers and failed to find a publisher (according to Howard, who had the manuscript before him when *he* wrote) because Jones, although trying

> to be fair and honest in his criticisms as he would in reviewing the work of a colleague . . . made the mistake—unconsciously, no doubt —of instilling into his otherwise impeccable statements slight traces

> of institutional venom. One publisher . . . offered to hold the manuscript and use it *after* Burbank's death. The author naturally refused this proposal as it would have defeated his purpose in writing . . . which was to dethrone a popular idol.[3]

Such was the atmosphere of controversy—by now so long dissipated that few people have any clear idea of who and what he was, let alone critical opinions—that surrounded the name Burbank during the twenties.

He remains a difficult subject for the biographer. Notwithstanding Howard's labors, even the list of Burbank's plant introductions—numbering between 800 and 1,000 over a working life of some fifty years—remains inconclusive. The work of the horticulturist may extend over decades, experiments taking many years to complete. Dates are frequently obscure or contradictory. And in any case, as Thoreau—a writer always close to Burbank's heart—observed in his *Journal*:

> In a true history or biography, of how little consequence those events of which so much is commonly made! For example, how difficult for a man to remember in what towns or houses he has lived, or when! Yet one of the first steps of his biographer will be to establish these facts, and he will thus give an undue importance to many of them. I find in my Journal that the most important events in my life, if recorded at all, are not dated.

In writing Burbank's biography, I have had access to two primary sources not available to Jones and Howard. These are the correspondence of Edward J. Wickson, for many years his friend and best promoter, and the papers of George H. Shull, one of the outstanding American plant geneticists of the day, who observed him at work on behalf of the Carnegie Institution of Washington from 1906 to 1911. Shull, too, had planned to write a book about Burbank, and no one was ever better qualified to do so, but it was never completed.

I have also, of course, had the use of Jones's manuscript (its author is, interestingly enough, linked with Shull as cooriginator of the pure line method of corn breeding, which revolutionized agriculture) and of Walter L. Howard's very extensive collection of what he called "Burbankia," much of which he was himself able to use only indirectly. Howard does not seem to have known of the existence of the Wickson papers. He was denied access to Shull's notes because the latter, though his project had been shelved, never entirely gave up the idea of bringing out a book of his own on the subject. "Shull," Howard says despondently, "could give us more information about Burbank than any man living, but he will not talk."

The essential elements of the Burbank story are thus brought together here for the first time. That it can be told at all is substantially owing to the work of Walter L. Howard, who was as tireless in researching Burbank's life as he was in cataloguing his plant introductions, and of George Shull, an observer in the best scientific tradition: this book is accordingly dedicated to their memory.

I am indebted to the director of the Bancroft Library of the University of California at Berkeley and to the Howard family for permission to use the materials assembled by Walter Howard; to the Connecticut Agricultural Experiment Station for permission to quote from Donald Jones's manuscript "The Life and Work of Luther Burbank"; to the Carnegie Institution of Washington, D.C., the American Philosophical Society, Philadelphia, and the Shull family for their assistance and the use of the papers of George Shull; and to the Library of the University of California at Davis for access to the correspondence of Edward Wickson. Unless otherwise indicated, quotations in the text attributed to Howard, Jones, Shull, and Wickson have their sources in these unpublished documents. Extracts from Burbank's posthumously published autobiography, *The Harvest of the Years*, written with Wilbur Hall, are quoted by kind permission of the Houghton Mifflin Company.

So many people have assisted in the making of this book that the list given here must necessarily be an abbreviated one. The advice and support of Peggy Brooks, the editor of the original edition, and of Peter Matson and Robert Briggs were invaluable. Preparation of the revised edition, based in part on the Luther Burbank materials deposited in the Library of Congress on the death of Elizabeth Burbank in 1977, was greatly aided by a grant from the Luther Burbank Museum, Santa Rosa. The confidence and generosity of the museum's advisory committee and, in particular, of its chairman, Sherman B. Boivin, are gratefully acknowledged.

This new edition has been so extensively rewritten as to be virtually a new book. I believe it to be both much more readable and more informative in numerous ways than the original. Revision was made possible in the first place by the benevolent judgment of William J. McClung and the editorial committee of the University of California Press. To the editors and staff of the press, old friends by now, and to all those others who helped, my enduring thanks.

1

THE SAGE OF SANTA ROSA

One July afternoon early this century, a remarkable expedition, including three world-renowned scientists, made its way to the modest Santa Rosa home of Luther Burbank. The local people were not surprised. Burbank was, after all, one of the best-known men in California—perhaps even in America—and curious folk, scientists included, often came to visit him. There was always someone wanting to talk about his work, to shake him by the hand and tell him of their admiration, or simply to catch a glimpse of him in his famous experimental gardens. The reputation of the great "plant wizard" had spread far and wide. That was easy enough to understand. Here, apparently, was a man who turned out wonderful new fruits, vegetables, and flowers in a seemingly unending stream. There were Burbank potatoes in every pot, Burbank plums at every table. And this, they said, was just the beginning, with rumor now of miracles such as the white blackberry, the stoneless prune, and a spineless cactus that would open up the deserts to stock raising. Farfetched gossip even told of oddities such as coreless apples, seedless tomatoes and melons, huskless corn, green carnations, and a black—or perhaps it was blue?—rose. No telling how he did it either. He was one of a kind. Most Santa Rosans were secretly a little in awe of him, though proud to be his neighbors. Single-handedly he had put their little town on the map. And for all that he was such a down-to-earth, regular sort of fellow. He had the common touch and genius besides. It was a combination that was hard to beat.

There was in him, too, more than a little of the "prophet of the soul" Emerson had spoken of. And, like Emerson, the public loved a prophet of the soul. Behind his back (despite his ready sense of humor, it did not amuse him), they probably repeated the old joke: "Luther Burbank crossed the milkweed with the eggplant and produced an omelet." Joking aside, though, they were impressed. He was not without honor in his own country. Small wonder, then, that even foreigners—and important foreigners at that—came all the way here to see him.

The leader of the visiting group, a big, heavily built man, a full head taller than Burbank, with graying hair and beard, was the great Dutch botanist Hugo de Vries, one of the three researchers who in 1900 had independently rediscovered and confirmed the laws of heredity so painstakingly worked out more than thirty years before by the Austrian monk Gregor Mendel (1822–84). His was a mind much in advance of its time. "If the stature of de Vries has steadily increased as genetics has developed," writes L. C. Dunn, "it is perhaps in part because genetics grew to resemble the view he entertained in the 1880s—that the variety of the organic world was maintained through the properties and activities of material units which were the prime operators in the transmission of heredity, in evolution, and in development."[1]

With De Vries—the most important member of the group from Burbank's point of view—were two other prominent scientists. Svante Arrhenius of Sweden was one of the founders of modern physical chemistry. The previous year, he had been awarded a Nobel Prize for his theory of electrolytic dissociation, which was based in part on De Vries's experimental observations. Jacques Loeb, a German-born biologist who was now a professor at Berkeley, was famous for his controversial experiments in artificial parthenogenesis. Loeb's work was also of considerable significance to the question of inheritance. In 1899 he had successfully caused sea urchin larvae to develop from unfertilized eggs by controlled changes in their environment, later going on to produce and raise parthenogenetic frogs to sexual maturity. Guiding the party were

Edward Wickson and W. J. V. Osterhout, two other professors from the University of California, both of them friends of Burbank's.

In addition to participating in the dedication of the Carnegie Institution's new Station for Experimental Evolution at Cold Spring Harbor, Long Island, and giving a course of lectures at the University of California, De Vries had come to the United States to observe the evening primrose *Oenothera lamarckiana* in its native habitat. This was the plant that had led him to the concept of independent factors in heredity and ultimately to Mendel and the theory of mutations.

But he was also intensely interested in the new varieties of fruit being produced in California. "One reason which, more than others, made me accept an invitation to visit California," he admitted, "was the prospect of making the personal acquaintance of Luther Burbank." If any man living could shed light on the manner in which new traits came about, it would, he felt, be Burbank. He had hoped to meet him at a congress of hybridologists in London a few years before, but the American had, as usual, advanced the demands of his work as an excuse for not attending. Now, at the busiest time of the year for the experimental hybridizer, De Vries feared that Burbank might have no time for visitors.

When it suited him, however, Burbank was prepared to make time. It was being dragged off his own ground, to congresses, conventions, and lecture halls, where he might—heaven forfend!—be called upon to speak, that he hated. And if the importunities of the ill-informed and interfering bored and irritated him, he was more than flattered by the attentions of so select a company as this. Here in Santa Rosa, surrounded by the visible achievements of almost thirty years' labor, he felt none of the diffidence that had plagued him all his life. Loeb, Arrhenius, and De Vries were names to conjure with, luminaries of the scientific firmament by any standard. His secret ambition was to be considered one of them—to be accepted into the great fraternity that had inherited the mantle of his boyhood hero, Darwin. With this meeting, it seemed as though the long-delayed approbation of the hierarchy of science was finally being granted. He was fifty-five. In his own field—the development and introduction of new varieties of useful plants—he was second to no one. He must have thought it none too soon.

For their part, the visitors found in their host an outwardly unassuming sort of man, short, spare, a little stoop-shouldered, but with sparkling blue eyes and a lively sense of fun. The house, where he lived with his mother, sister, and one servant, was small and unpretentious. Burbank talked eagerly, regaling them with anecdotes that kept them laughing and taking down the photographs of his triumphs that covered the walls to explain them. He had more of the gardener about him than the savant, De Vries thought. Perhaps the Hollander even wondered, as others before him had done, if this could actually be *the* Luther Burbank who had accomplished so many wonderful things and earned himself such a reputation.

They spoke of Burbank's continuing work in crossing the wild beach plum—a scrubby tree able to grow and fruit under the most adverse conditions—with other American and Japanese plums. By this means, hybrids had been created that combined the vitality of the beach plum (the fruit of which was almost worthless except for making preserves) with size, flavor, and overall market quality. This involved not merely crossing six or eight types with one another but using as many subspecies and varieties as possible on a mass scale. And the only proof of the worth of each crossing—from which new combinations were indicated—was the fruit itself. "It can easily be seen what an immense amount of work, patience and capacity of judgment and choice is required to reach the ultimate aim," De Vries observed. "Yet Burbank told us on that remarkable evening of many such instances."[2]

The three Europeans were eager, above all, to discuss the theoretical implications of their host's work. De Vries and Loeb, whose own efforts had been directed at pinning down the nature and origin of new evolutionary traits, were particularly anxious to know whether Burbank could cast any light on this, and De Vries took the first opportunity to bring the conversation around to the subject. One of the marvels they had seen at Burbank's Sebastopol experiment grounds near Santa Rosa was the so-called "stoneless prune." This was a small, blue-colored plum with a seed something like an almond (it even tasted like an almond), surrounded by a pale, jellylike integument, with only a few remnants of stony material, which could easily be bitten into. ("To take up a plum and bite through it without hesitation requires education," noted

Wickson, "so strong is the conception of the danger involved; but to bite freely and find the flavor enhanced by the nutty savor of the kernel brings reward in the new sensation which the palate experiences. . . . To combine the flavors of pulp and kernel, to gain the nutritive properties of the latter and to escape the tedium and awkwardness of ejecting the stone, constitute an advance in prune character and motive which it is difficult to overvalue."[3] In some of the hybrids, more of the stony material survived, in others less. Burbank's plan was to cross the latter with superior varieties of prune in order to transmit the stoneless quality to them. Was this fruit evidence that a major trait such as stonelessness could be arrived at by crossing ordinary varieties?

No, said Burbank. He had heard of a *prune sans noyau* (seedless plum) known as a curiosity in France since at least the sixteenth century, and to be found there now as a worthless wild fruit of the hedgerows. Seeing the potential use of this oddity in hybridization, he had sent to a French nurseryman for the seed. This was grown at the Sebastopol grounds and eventually produced a fruit about the size of a small cherry, with a damson color and flavor and a stone partially covering the pit, growing on an unproductive, rambling, thorny bush. After repeated crossings with the standard French prune and other plums, the quality of the fruit was slowly improved, while the remnant of stone was kept as small as possible. After about ten years' work, sweet and sour damsonlike fruit of various sizes resulted. There was, therefore, no exception to the rule: no really new character.

The visitors were a little disappointed. They had hoped for some novel revelation. Even they (and for De Vries and Loeb, we should remember, this problem was "the fundamental idea, if not ultimate aim" of their studies) seem to have succumbed to Burbank's almost mystical reputation. Like so many of his lay admirers, they expected too much, experts of international stature though they were.

The other extraordinary things Burbank had to show them, impressive though they might be, also failed to clarify the fundamental theoretical question. There was the white blackberry, for example, a hybrid that produced an abundance of perfectly white fruit. (Delicious, De Vries thought.) How could so basic a feature as the black color of the berry be eliminated merely by crossing?

Again, there was no magical solution, simply the ingenuity of the breeder. A white variety of raspberry was known to occur in Europe, Burbank reminded them, and a cultivated bramble with brownish yellow berries of inferior quality was found in the eastern United States. He had crossed this last with a variety known as Lawton's blackberry, interbred selected progeny, and in the end had been rewarded with a berry that was an almost translucent white and had the eating qualities of the Lawton.

The same was true of the "spineless" cactus by which Burbank set so much store; he had assiduously collected varieties of cactus from Mexico, South Africa, and other countries* until one finally turned up that was without the usual spines on the stalks, and another that lacked spicules on the pads. These characteristics were combined in a single plant by hybridization after an extensive series of crossings, and a spineless cactus that its creator saw as having immense potential as cattle forage in desert areas was produced. Now and then a spine still occurred on the stems, but after a careful search De Vries was able to find only one. Burbank demonstrated the harmlessness of his cactus by softly rubbing his cheek against the pads. It was a remarkable achievement. But it was no miracle.

What, then, was the secret of his success? First of all, there was the sheer scale of his experiments, which no other plant breeder could match. In this sense, he was the Henry Ford of the art: he brought mass production to hybridizing, raising thousands of seedlings to obtain a single improved variety. This naturally increased the chances of finding useful specimens, inasmuch as the likelihood of discovering a valuable "sport" among thousands of plants was obviously better than among hundreds. And since the chances were that out of five or six desired qualities, only three or four ordinarily appeared in combination, thousands of seedlings had to be produced to find one that would incorporate them all. In a single year, Wickson observed, fifteen huge bonfires were lit on Burbank's grounds to destroy rejected material, one of them alone burning 65,000 two- and three-year-old hybrid seedling berry bushes. As it was just after fruiting time that the rejected speci-

* Almost all cacti were originally native to the New World. Burbank was, in effect, reimporting varieties naturalized elsewhere in the hope of finding variants resulting from selection in different environments.

mens were destroyed, it was hardly surprising that his neighbors wondered at "the man who used to have a big nursery, but now raises acres and acres of stuff and every summer has it all dug up and burned."

There were hundreds of stock trees on his place, each of them grafted with dozens of different hybrid plums and covered with an amazing variety of fruit and foliage. No doubt but that he had to be given credit for demonstrating the possibilities of breeding on a large scale. From the perspective of the present day, it is in this that he stands between the small-scale farmer-breeders who through most of history slowly but steadily built up mankind's stock of useful plants and the modern scientific hybridizer. De Vries was quick to recognize this.

Another factor De Vries conceded was that Burbank seemed to be guided by a "special gift of judgment, in which he excels all his contemporaries." That was high praise indeed, coming from a man who had studied the work of plant breeders all over Europe, including that of the great cereal hybridizer Hjalmar Nilsson, director of the Swedish Agricultural Experiment Station at Svalöf. This "gift of judgment" was no insignificant part of Burbank's secret. Without it, De Vries felt, his imitators were bound to fail. No one could easily "steal his trade," for "without the special disposition for it nobody will succeed, and for simple imitation, the entire process is too complicated."

Apart from this and the scale of his experimentation, Burbank's techniques were essentially the same as those used by plant breeders in Europe. But since the procedures and results of European horticulturists were little known in America at this time, he had had to rediscover many of the common European methods on his own—in itself no small achievement.

De Vries freely recognized that in the number of fruits and flowers Burbank had improved, he had no equal. Where other breeders had tackled a few genera, usually as a business proposition, Burbank had taken for his province the whole catalogue of cultivated plants and had abandoned the seedsman's trade so as to devote himself exclusively to the work of improvement.

The Hollander had some reservations, however, which he voiced to a young Dutch gardener named John Zuur who worked for Burbank. Zuur understood English but could not speak it well, and he was naturally pleased to be able to talk to his famous compatriot in his own language. He told De Vries that he found Burbank eccentric and suspicious of his workers. The men were not allowed to talk to visitors, and if Burbank caught one of them talking to someone over the fence, he was likely to be fired—not for wasting time but for talking about Burbank's business. Burbank *was* afraid that people would steal his secrets. To prevent seeds, bulbs, and scions from being stolen, it was his practice to search the men's pockets as they left work. "He treated us as 'high-graders,' one former employee complained, using a term applied to workers in the goldfields who stole bits of high-grade ore. And one could never tell when Burbank might do or say some unexpected thing. "See here, Zuur," he had remarked out of the blue one day, "I think I am about 200 percent overrated!"

For his part, De Vries told Zuur that Burbank's failure to keep proper records of his experiments was a disappointment. He also expressed regret at Burbank's inability to produce entirely new traits. The first criticism was destined to be voiced repeatedly in the years to come. It was valid enough: Burbank's records were usually intelligible only to himself, if then, and vital details of the ancestry of his hybrids were frequently lost or uncertain. The second was unfair. After all, no one else had ever successfully produced new traits by breeding. Selection and hybridization could only combine and enhance existing characteristics, and mutation was a factor then beyond human influence.

As we have seen, however, even scientists tended to expect miracles of Burbank and were disgruntled when he failed to come up with them. Academic training and distinction do not necessarily make for superhuman objectivity. De Vries had behind him the experience of finding in Mendel an amateur observer who, almost unknown to the world, had arrived at a solution to the problem of inheritance that had gone unnoticed by formal biology. By virtue of his part in the rediscovery of Mendelian law and his own theory of mutation, the Hollander was finding himself hailed as a second Darwin. But there were many who rejected the theoretical edifice he had constructed, among them the venerable Alfred Russel Wallace (1823–1913). Wallace—whose authority was that of joint originator of the theory of evolution by natural selection and whose independent findings had pushed Darwin to publish in the first place—stubbornly held out against Mendelism. So did many other scientists. And the theory of dramatic mutative leaps proposed by De Vries was even more contentious. Its author was, in other words, a man with an ax to grind.

Perhaps De Vries had secretly hoped to discover in Burbank another Mendel, who would produce confirming evidence of massive mutations as the source of new species. At very least, he must have hoped to find materials in Burbank's collection that would illuminate his theory. Burbank, De Vries declared, was "a gardener touched with genius"—and was that not precisely what Mendel had been? Like Mendel, who at least once failed to pass an examination in natural history, Burbank had scant academic training. To the hopeful De Vries, the parallels with the saintly Augustinian abbot must have seemed remarkable. It must have been tempting to suppose that answers to the complex problem of how species arose and differentiated could be had for the asking from the sage of Santa Rosa.

But De Vries was doomed to disappointment on that score. On the evidence, in fact, Burbank probably jolted his confidence in his own mutation theory.

When De Vries and company visited him, Burbank was already world famous. The obloquy that was later to cloud his name, particularly in scientific and horticultural circles, was still in the future. He pursued his goals with an innocent fervor, as yet unembittered by the experience of commercial exploitation and scientific disdain. But the tide of popular adulation that was ultimately to swamp him was rising.

He worked exceedingly hard. Now, in July, the busiest time of the year for the experimental horticulturist, when fruit was ripening, when maturing seeds had to be collected, dried, and stored, when the selection of plums took place, he would rise at dawn, and by 5:00 A.M. would customarily be starting his rounds in the Sebastopol experiment grounds, seven miles from his home, sampling fruit, judging it for size, shape, color, and taste, and recording his observations.

July, Walter L. Howard observes, was also the month of conventions, and "a state or national convention in San Francisco was never complete unless the delegates could spend a day in Santa

Rosa seeing the Burbank gardens. Travel bureaus and chambers of commerce encouraged this."[4] Burbank seldom delegated authority where the work of selection was concerned, and his public, for its part, wanted to see *him*, not just some spokesman. Howard paints a revealing picture of Burbank returning home from his labors in the hot sun at Sebastopol, his stomach soured by sampling too many acid fruits, to face the penalties of celebrity:

> Passing from the garage to the house he runs the gauntlet of outstretched hands and cheery greetings. He bows right and left and impatiently tells the callers he regrets that he cannot stop to talk with them. At the door a man waylays him and grabs his hand only to be thrust aside. Another more daring than the rest follows him into his study and insists upon introducing himself and explaining why he should have an interview. He is asked to leave.
>
> After Burbank finally sits down to lunch the telephone announces that a party of seventy-five or a hundred persons have arrived in town and wish to be conducted over his gardens. The local Chamber of Commerce secretary protests that the party was sent over by a travel bureau, a plan Burbank had approved months earlier, and what should he do with them. Burbank capitulates and the party comes, but he is not a gracious host. Some were grateful for having seen him under any circumstances; others considered him to be peevish; while the few that had forced their attentions upon him and were repulsed said he was rude.[5]

Fame, which Burbank certainly enjoyed for its own sake and had in his way courted, was proving a harsh mistress. We can see him now as an early example of that unfortunate species, the media hero, suffering all the penalties attached to the role. But what was it, after all, that had made him so famous? What was it about him that appealed so profoundly to public fancy? He had become one of the undoubted idols of his time, an age that led directly to our own, and answering these questions should tell us a good deal about the essential temper by which the imagination of the twentieth century was delineated.

There was in Burbank that Emersonian faculty of prophecy, and what he prophesied was a land of Canaan where blackberries were as big as plums and plums the size of peaches, where everything grew straight and easy and the pastoral republic that had vanished in living memory was restored to a people sickening of the spreading fumes and ashes of the Industrial Revolution. The railroads were all built and all the free land was taken. The star of empire could go no farther west. Henceforth, the American frontier would exist only in dreams, and Burbank was a spokesman of the dream. Perhaps he speaks to us still, in the eternal longing of urban man to reconstitute the Garden of Eden (more commonly thought of as a little place in the country) one day.

2

THE ESSENTIAL ART

The story of Luther Burbank belongs to the history of horticulture—the science and art of growing plants and of their selection and improvement for human purposes. Of all the relations of people to their world, there is none more basic than this. Soldiers and kings, politicians and philosophers have claimed the lion's share of attention, but peasant farmers have underwritten the history of humanity in the unending process of cultivating food and culling seed for the next season's planting. Though people recognizably like ourselves may have been on earth for as long as a quarter of a million years, the pattern of life remained relatively static until the invention of agriculture a bare ten to thirteen millennia ago. In effect, history begins then.

In the process of domesticating plants and animals, man also domesticated himself. Cities became possible, springing quite literally from the development of the granary. Metallurgy, weaving, pottery making, the basic mechanical principles, the plow, and the wheel made their appearance. In the transformation and improvement of cultivated plants, we may trace the genesis of civilizations. We have depended on them absolutely ever since the appearance of the first substantial villages in Asia Minor and Middle America heralded the beginnings of urban life. Wheat, rice, barley, oats, rye, corn, sorghum, and millet: one or more of these has lain behind every great increase in human population. Even now society as we know it would be unlikely to survive without them.

Where other records fail, cultivated plants are, in the words of the geographer Carl O. Sauer, "living artifacts of times past," from which careful detective work can deduce the unwritten history of mankind. The difficulty lies in reconstructing their heritage and translating their message, which faithfully preserves the labor and achievements of the experimenters and cultivators who laid the foundations of our society and culture. In this record, it is not unreasonable to surmise, may lie lessons of importance in the coming struggle against ecological catastrophe.

But though civilization, science, and empire have from the very beginning followed the plant breeder who could make two blades of grass grow where one grew before, no one was ever more anonymous. "Our civilization still rests, and will continue to rest, on the discoveries made by peoples for the most part unknown to history," Sauer observes. "Historic man has added no plant or animal of major importance to the domesticated forms on which he depends."[1] Even the safflower, hailed as a "new" crop and now raised on many hundreds of thousands of acres in the United States for its oil, has been found in Egyptian tombs dating back to 1500 B.C. and was cultivated then for the dye obtainable from its flowers.

In historic times, too, the work of horticulture has been accomplished with little fanfare. Luther Burbank was the first, and is thus far the only, plant breeder to have his name become a household word, bestowed with little discrimination on high schools and reformatories, and written for a time into *Webster's* dictionary as a transitive verb:

> burbank, *v.t.* To modify and improve (plants or animals), esp. by selective breeding. Also, to cross or graft (a plant). Hence, figuratively, to improve (anything, as a process or institution) by selecting good features and rejecting bad, or by adding good features.

The most probable candidate for this kind of fame today is perhaps Norman Borlaug, father of the "green revolution" that has revolutionized agriculture in many countries. But Borlaug, notwithstanding his outstanding achievements and Nobel Prize, shows no signs of becoming the object of a cult, venerated by millions and accorded an almost superstitious reverence as was Burbank during his lifetime.

Even for Burbank, however, fame has proved relatively fleeting. The verb failed to endure. There is a tendency among Californians to assume mistakenly that he was the eponymous founder of the city of Burbank.* Oddly enough, many confuse him with the black agricultural experimenter George Washington Carver (1864–1943). Perhaps only in California's Sonoma County, his workplace for some fifty years, do most people still know who Luther Burbank was.

It may be argued, of course, that Burbank was more celebrated in the first place than he ever deserved to be. Orthodox scientists certainly tended to think so. (To them he might have replied in the words he used in a lecture at the University of California's Hearst Hall in 1908: "Orthodoxy is ankylosis—nobody at home; ring up the undertaker for further information.") Scientists found his failure to keep their kind of records irritating, and many of them eventually came to think him a fraud. In the end, they damned him by silence, a technique all too frequently employed to deal with those who cannot easily be boxed into the conventions of the day. Burbank was to start with an outsider with scant formal education. His achievements were arguably remarkable, but he was guilty of failing to maintain a properly scientific facelessness. Even though it had never been his intention to do so, scientists did not thank him for turning horticulture into a kind of spectator sport.

The public streamed to his door in their thousands. There was a natural constituency of Burbank enthusiasts, and it was not only in the United States. Henry Ford, Thomas Alva Edison, Helen Keller, Elbert Hubbard, William Jennings Bryan, Jack London, John Burroughs, William Howard Taft, John Muir, Sir Harry Lauder, Paderewski, and the king and queen of the Belgians† were numbered among his famous visitors. Something about him drew a responsive chord. He possessed the mysterious quality that goes by the name of charisma. His name, observed his Indian friend Paramahansa Yogananda, became a synonym for improvement. And improvement, or "progress," was the tutelary genius of the age.

* It was named for a Dr. David Burbank, originally of New Hampshire, who operated a sheep ranch at the site from 1867 and joined in founding the town twenty years later.

† The Belgian royal couple seem to have had an enthusiasm for sages, which also led them to cultivate the friendship of Albert Einstein.

In order to understand what it was that Burbank did, it is necessary to know a little of the history of the crucial symbiosis of man and cultivated food plants. The long process of bringing favored species into domestication involved more than 10,000 years of painstaking selection, largely carried out by unknown experimenters. The Massachusetts fields and gardens in which Burbank began his lifework came stocked with corn, beans, and squash that were the products of Central American farmers, potatoes and tomatoes originating in the empire of the Incas, spinach, carrots, and peas from Iran and Afghanistan, apples and onions from somewhere in western Asia, cabbages, turnips, and plums brought into cultivation in neolithic Europe, cucumbers from India, and peaches from China, to name only a small part of their contents.

Roots and tubers were probably eaten raw for a very long time, but the human gut does not digest uncooked starches and vegetable protein well, and our ancestors quite early gained control of what was to be the most useful of all their tools: fire. We know, for example, that fires burned in the famous cave at Choukoutien, near Beijing, inhabited by Sinanthropus a quarter of a million years ago. The discovery of cooking was not the only consequence. Millennia before the beginnings of agriculture, humans set fires, deliberately or accidentally, that drastically altered the plant covering of the planet and had the effect of encouraging the variation and development of some species at the expense of others. People also involuntarily influenced plant life in other ways. Human habitation sites were a novel environmental niche, providing the opportunity for new varieties of plant to grow and adapt. We commenced to exert a selective influence on the vegetable world long before we knew what we were doing. A relationship sprang up between the useful weeds that throve in a human context and the hominids who gathered their fruits and flowers, roots and stalks, for food, fuel, and other purposes. The instincts of the earliest humans, and even, perhaps, of the primates who preceded them, derived from hundreds of thousands of years of experience, were perpetuated in unbroken continuity as an increasingly intimate understanding of the inherited food plants.

These processes have gone on for so long that it is scarcely meaningful to speak of a natural balance without man. The dim beginnings of agriculture itself are, however, a matter of debate. It seems probable that it commenced somewhere between 10,000 and 13,000 years ago, both in the Old World and in the New, at the late glacial/early postglacial time boundary. In an era of intensified food collecting, called the Mesolithic in the Old World and the Archaic in America, perhaps partly as a consequence of increasing populations, actual cultivation began. Systematic irrigation of wild plants may have preceded it.

Plant domestication took two basic forms: vegetative reproduction, or cloning, by planting a segment of the parent, as in the case of the potato and the banana, and seed planting, as in the case of wheat and beans. Carl Sauer has argued that the second arose out of the first. According to this hypothesis, the earliest plantings would very likely have been for purposes other than food— perhaps for the extraction of poisons or narcotics used in stunning or killing fish. "It may well be," Sauer writes, "that among the earliest domesticates were multi-purpose plants set out around fishing villages to provide starch food, substances for toughening nets and lines and making them water resistant, drugs, and poisons."[2]

With the selection of varieties for purposes of planting came an increase in extractive technology. Some foods had to be processed before cooking; new food combinations were experimented with. Botany and chemistry—and hence science itself—can in some sense be seen as arising out of these early experiments in farming and eating.

The growing of crops by seed planting may have arisen almost accidentally, Sauer has suggested, in the utilization of "volunteer" weeds that sprang up among the cloned species. Thus rice is thought by some authorities to have originally been a weed in fields of taro, a plant of the arum family grown for its roots, in Indonesia or India. As any gardener will have observed, the planting of crops provides room for self-sown plants. Wild varieties of tomato that are protected but not planted deliberately are, for example, still gathered in parts of Mexico and Central America.

As time went by, certain of these "weeds" would themselves come to be planted deliberately, and as the "Old Planting" people envisaged by Sauer gradually expanded into areas less favorable to their original cloned plants and more so to the seeded species, the latter would be increasingly utilized and at the same time selected and improved.

An alternative hypothesis is that seed planting developed from the collection of wild grain. In 1966 the agronomist J. R. Harlan made the experiment of harvesting wild wheat in eastern Turkey using a 9,000-year-old flint sickle set in a new wooden handle. He found he was able to gather a kilogram of grain an hour. On this basis a family working for three weeks when the wild grain was ripe could have gathered more than they could consume in a year. Harlan's grain was, moreover, found to contain at least 50 percent more protein than modern American bread wheat. In this context Kent V. Flannery, of the University of Michigan, has proposed a period of "preadaptation" when diminished food supplies, specifically in the Near East, but perhaps elsewhere too, laid the basis for the deliberate cultivation of food plants and the domestication of animals.

In the Old World, seed-planting agriculture may conceivably have begun before the end of the most recent Ice Age, in areas now desert and hard to envisage as fertile. During the Ice Age the pressure of cold air over Europe caused the Atlantic rains to follow a more southerly route to the east, and a belt of wood- and grassland covered the whole area from the west coast of Africa to the mountains of Iran. Old World centers of seed domestication are believed to have been Asia Minor and western India, northern China and Ethiopia. (Another area of early cultivation may have been the great bend of the Niger River in West Africa.) In the absence of reliable archaeological evidence, however, the pinpointing of the original hearths of domestication remains a matter of dispute.

The earliest known seed planting took place in the hills flanking Mesopotamia's "fertile crescent." Here the archaeological documentation exists, brought to light by excavations in what is now Kurdistan (forming part of Turkey, south of Lake Van, Iran, and Iraq), and in Palestine, where the wild ancestors of wheat and barley still survive. Here, as long ago as 7000 B.C., village farming communities existed, cultivating wheat, barley, field peas, lentils, and other leguminous plants, and collecting nuts, fruit, and berries growing wild in the area. The process of plant improvement proceeded relatively rapidly. Already by the seventh millennium B.C., wheat and barley in particular had come a long way from their wild ancestors.

The seed-planting culture originating in the hill country moved down the river valleys to the alluvial plain of the Tigris and the Euphrates. Now, for the first time in history, we find people in sufficient numbers and with sufficient control of their food supply to build cities. Somewhere around 4000 B.C., with the rise of the Sumerian city-state, civilization as we know it begins.

Agriculture would appear to have reached Europe via Anatolia some 2,000 years earlier. By the time the first Sumerian cities were being built, it had already arrived in the lands bordering the shores of the North Sea. Emmer wheat and six-rowed barley, originally brought into cultivation in Asia Minor, were carried into the Balkans by neolithic farming peoples who tilled temporary clearings in the forest and moved to a new site when the fertility of the soil was exhausted. As always, the process of hybridization and selection automatically accompanied the movement of populations. The cultivated cereals, in the process of being introduced into environments differing from those in which they had originally been domesticated, slowly changed their nature. No conscious human intervention was necessary.

As these peoples moved north and west, whether as a result of population pressures, the vicissitudes of their slash-and-burn planting, or out of sheer adventure, the nature of their crops inexorably altered to exploit the potential of climate and soil. Natural selection continued to operate, though man was now one of nature's agents. Rye and oats, originating as grain-field weeds in Asia Minor, and perhaps accidentally imported into Europe in the seed of wheat and barley, came into their own in the harsh climate of the north. Animal husbandry became an increasingly vital part of the rural economy, necessitating the cultivation of special fodder. "Such were the systems of culture across northwestern Europe for two to three thousand years, systems not seriously modified until the eighteenth century," Sauer says. "Then, with the introduction of the potato, the development of stock beets and field turnips, and the cultivation of clovers, the new agricultural revolution arose and, in part, prepared the way for the industrial revolution."[3]

But in the several dozen centuries that intervened, horticulture had not stood still. Greek and Phoenician colonizers and traders, ranging from the Black Sea to Gibraltar and beyond, greatly promoted the spread of plant varieties and agricultural technology from the eastern Mediterranean. The Roman writer Varro lists over fifty Greek authors who wrote on agriculture and related topics; one of these, Theophrastus, is considered to be the founder of systematic botany. The Carthaginians, drawing on Greek sources for technical information, developed a system of intensive plantation farming of crops such as the olive and the vine. Mago of Carthage, called the "father of husbandry," wrote a lengthy work on agriculture in Punic, which was translated into Latin at the orders of the Roman senate shortly after the fall of Carthage in 146 B.C. Rome and her empire were themselves supplied by an elaborate agricultural substructure, which has been studied in detail.[4] Much of what was known was preserved into the Middle Ages.

European agricultural technology may be presumed to have stagnated during the declining years of the West Roman Empire, but there was a wave of expansion of land under cultivation after A.D. 600. The introduction of the three-field system of crop rotation under Charlemagne and a variety of new inventions—the horseshoe, the heavy Saxon wheeled plow, modern collar harness, and improved wagons equipped with brakes, whiffletrees, and pivoted front axles—to a great extent underlay the high civilization of the Renaissance and the rise of modern Europe.[5]

The agricultural revolution of the eighteenth century, based both on new food plants and on better methods of cultivation, followed in its turn. The English agriculturalist Jethro Tull (1674–1741) invented a drill for planting seeds in rows, and in his book *Horse-Hoeing Husbandry* advocated manuring, crop rotation, and repeated tilling to eliminate weeds. Agricultural experimentation became fashionable among the great English landowners. For hundreds of years, however, there had been a slow, steady improvement in European cereal yields. In England, for example, the average return during the period 1200–49 was 3.7 grains harvested for each grain sown; by 1500–1700, it was 7 for 1; and by 1750–1820, it had reached 10.6 for 1.[6] These impressive gains may have been the result of better varieties. Systematic improvement of cereals by selection seems to have begun only in the early nineteenth century, but it had been known at least since Roman times (Virgil alludes to the fact) that care had to be taken in selecting grains destined for sowing, or the races would deteriorate. The peasants slowly and silently carried on this work. Science would be a while catching up with them.

Basic to hybridization—the mating together of related varieties and species to produce new types—is an understanding of sexuality in plants. There is evidence that this knowledge existed in the ancient world. Herodotus, writing in the fifth century B.C., remarks that the Babylonians cultivated the date palm as they did figs, bringing the flower clusters of the male trees from the desert and binding them to the fruit-bearing trees, to which alone they gave a place in their gardens. The significance of doing so does not seem to have been properly understood. It was believed that insects from the male flowers themselves fertilized the fruit-bearing trees rather than the pollen they carried.

The modern study of sexuality in plants began in 1676 when Nehemiah Grew proposed to the Royal Society in London that the stamens are the male organs of plants and pollen acts as vegetable sperm. Experimental proof that pollen was necessary for seed development in a number of plants, among them *Zea mays* (corn), was provided by the German botanist and physician Rudolph Jakob Camerarius in his *De sexu plantarum*, published in 1694. In 1716 the first American to be made a fellow of the Royal Society, Cotton Mather, reported observations on *Zea mays* including wind pollination, variety cross, and the resemblances of some of the progeny to the male parent. The first reliably recorded hybridization was made the following year. This was Fairchild's Sweet William, a cross between the carnation (*Dianthus caryophyllus*) and the sweet william (*Dianthus barbatus*) made in 1717 by the English gardener Thomas Fairchild, who noted that the progeny resembled both parents.

The binominal nomenclature and system of taxonomy employed in modern botany were introduced by the Swede Carl von Linné, known as Linnaeus (1707–78). In 1757 Linnaeus made what has been termed the first scientifically produced interspecific hybridization (between the yellow- and violet-flowered goatsbeards, giving rise to purple flowers that were yellow at the base). The great systematic hybridologist of the eighteenth century was, however, the German botanist Josef Gottlieb Kölreuter, (1733–1806), who made his first cross, involving two tobaccos, in 1760. Kölreuter went on to perform hundreds of others, and also wrote a number of important papers on sexuality in plants, hybrids, and hybridization.

There is, however, literary evidence for supposing that hybridization was practiced by gardeners much earlier. In George Puttenham's *Arte of English Poesie* (1589), for example, we find it observed that

> the Gardiner by his arte will . . . embellish [a flower or fruit] in vertue, shape, odour and taste, that nature of her selfe woulde never haue done: as to make the single gillifloure, or marigold, or daisie, double: and the white rose, redde, yellow, or carnation . . . the cunning gardiner . . . vsing nature as a coadiutor, furders her conclusions & many times makes her effectes more absolute and straunge.[7]

And in Shakespeare's *The Winter's Tale*, probably written in 1611, Perdita remarks that she does not plant "streak'd gillyvors [i.e., gillyflowers: carnations or pinks of the genus *Dianthus*], Which some call nature's bastards" in her garden because she has

> heard it said
> There is an art which, in their piedness, shares
> With great creating nature.

And Polixenes responds:

> Say there be;
> Yet nature is made better by no mean
> But nature makes that mean: so, over that art,
> Which you say adds to nature, is an art
> That nature makes. You see, sweet maid, we marry
> A gentler scion to the wildest stock,
> And make conceive a bark of baser kind
> By bud of nobler race. This is an art
> Which does mend nature—change it rather—but
> The art itself is nature.
>
> [4.4.82–96]

It is perhaps surprising in the circumstances that in a work on horticulture such as John Parkinson's *Paradisi in sole paradisus terrestris*, published in 1629, there is no mention of hybridizing techniques. The seventeenth century undoubtedly exhibits a rising popular interest in systematic plant breeding and its products, however. During the "tulipomania" that raged in Holland in the 1630s, which subsequently provided the subject of Alexandre Dumas *père*'s novel *The Black Tulip*, fabulous prices were paid for new varieties. And the analogous "gooseberry fancy" that engaged the weavers of Lancashire in the years following 1650 apparently resulted in hundreds of new types.

U. P. Hedrick dates the beginning of agriculture by European settlers in New England to March 7, 1621, when, following their first terrible winter in the New World, the Pilgrims planted "garden seeds."[8] In Mexico and Peru, the Spanish conquistadores had long previously marveled at such indigenous novelties as avocados, guavas, cacti, maize, tobacco, tomatoes, potatoes, pumpkins, squashes, and new varieties of cotton, beans, and peppers. These American plants were to transform the everyday lives of people all over the world. Columbus little knew the significance of the discovery he made on November 5, 1492, when two of his men returned from the interior of Cuba with news of "a sort of grain they call maiz which was well tasted, bak'd, dry'd and made into flour." Maize, or corn, is today one of the most important plants in the United States and vitally significant to the agriculture of many other countries besides.

The northern settlements also found the scope of their farming greatly expanded by New World novelties. "North America is a natural orchard," Hedrick observes. "More than two hundred species of tree, bush, vine, and small fruits were in common use by the Indians when the Whites came. Besides these, there were at least fifty varieties of nuts, and an even greater number of herbaceous plants."

Among other fruits, the Indians had raspberries, blackberries, strawberries, blueberries, cranberries, mulberries, persimmons, plums, crab apples, cherries, and grapes. Some of these were deliberately planted and cultivated, others simply harvested from the wilds. The Cherokees and Iroquois are generally credited with having been the best husbandmen. Their horticulture was not entirely utilitarian either. There is some evidence that the Indians grew flowers in the vicinity of their vegetable gardens and orchards.

The main food crops of these people were corn, beans, and squash. Grown together as a complex, these constituted the agricultural triumvirate around which settled life revolved. Melons, sunflowers, and tobacco were, however, also grown in some quantity. It remains a curious fact that the North American Indians failed to cultivate root crops—though these were to hand and they dug wild tubers and roots (particularly the Jerusalem artichoke) for food. European settlers introduced the potato as a cultivated species in North America.

The colonists in the New World brought with them the skills and defects of their European motherlands. As agriculturists and gardeners, the Dutch, Huguenot French, Germans, and Swedes were preeminent. The main trouble with American agriculture in its early days was summed up by George Washington in 1791, in a letter to Arthur Young:

> The aim of the farmers in this country, if they can be called farmers, is, not to make the most they can from the land, which is, or has been cheap, but the most of the labour, which is dear; the consequence of which has been, much ground has been scratched over and none cultivated and improved as it ought to have been: whereas a farmer in England, where land is dear, and labour cheap, finds it his interest to improve and cultivate highly, that he may reap large crops from a small quantity of ground.[9]

For this reason, also, systematic improvement of food plants by selection was not much attempted in America until the end of the eighteenth century. Hybridization and the production of new varieties were even more restricted. "A study of records," says Hedrick (who certainly knew them), "would show that only an occasional cabbage, onion, lettuce, turnip, cauliflower, parsnip, or carrot was originated in this country before 1860." More, he admits, was done in the way of breeding flowers.

Nevertheless, it is possible to compile a lengthy list of Americans who applied themselves, with varying degrees of success, to producing new and better varieties in the late eighteenth century and the first half of the nineteenth century. Most of these men were denied the popular recognition their work merited. The public, unaware of what could be done in the way of improving plant varieties, also largely failed to notice what was being done.[10]

Among vegetable breeders, we might consider the work of the Ohio seedsman A. W. Livingston, whose efforts to obtain a better tomato led to the Paragon, introduced in 1870. Prior to that time there had been little interest in tomato culture, and the tomatoes grown lacked the desirable qualities of smoothness and uniformity. Tomatoes had been canned for sale since the 1840s, but up to the introduction of Livingston's Paragon, they had not been an important crop. His description of his procedure gives some idea of the state of the art in those days:

> I tried the best kinds then known to the public, and selected from these such specimen tomatoes as approached in qualities what was needed, or was in demand. The seed from these were carefully saved, and when planted were given the best cultivation possible, hoping in this way to attain what I desired. After 15 years of the most scrupulous care and labor of this kind, I was no nearer the goal than when I started in the race. According to the laws of life now (1893) well known, but which I did not then (about 1866) understand, such stock-seed would reproduce every trace of their ancestry, viz., thin-fleshed, rough and undesirable fruits. I ran this method through all its changes.

Some improvements were achieved, but these could be attributed to improved conditions, Livingston thought. Nonetheless, he found it worthwhile to put the better kinds on the market under various names, while he went on with the work, growing a varied selection of plants and keeping a watchful eye on his fields for any "leadings." Success followed on what he called his "new method"—selection of particular plants rather than of individual tomatoes. "Whether this method . . . was new to others at that time (in the 60's) I did not know," he says, "but it was altogether new to me; in fact it was a pure discovery on my part." He noticed a plant bearing heavy foliage and uniform, smooth tomatoes growing prolifically, but too small to have much market value.

> The seeds from this plant were saved with painstaking care, and made the basis of future experiments. The next spring, from these seeds, I set two rows across my garden—about 40 rods long each—and to my glad surprise they all bore perfect tomatoes like the parent vine. . . . They were a little larger, for which I also rejoiced. . . . The seeds from this crop were again carefully harvested, but from the first ripe and best specimens I selected stock to my own planting. By good cultivation and wise selection from season to season, not to exceed five years, it took on flesh, size and improved qualities. I then put it on the market. This was in 1870. . . . I called it the Paragon tomato.

Livingston believed he could repeat the process and was confident that he had produced an original distinct variety that would not deteriorate or "run out," for "they are as capable of being cultivated into 'strains' as are . . . cattle, hogs, chickens, or other plants and fruits of distinct kinds. The same laws of life and breeding govern tomatoes as in any other form of life, for all the processes of nature are so simple that few will believe them, even when they are pointed out to them." He later went on to produce a purple tomato—for which there was a demand in the western states—by selection. This was introduced in 1875 under the name Acme.[11]

Significantly, Livingston notes the absence of any kind of patent protection for the introducer of new plant varieties. Presumably, he did not get as much profit from his new tomatoes as he had hoped. An experienced grower, working only a few years before Burbank embarked on his career, he devoted over twenty years to the improvement of this one species. Any hybridization that was involved appears to have been accidental.

In these same years, however, the foundations were being laid for the development of scientific agriculture in the United States. The Morrill Land Grant Act of 1862 appropriated funds from the sale of public land to support agricultural colleges in every state, and the Hatch Act of 1887 established the agricultural experiment stations. In 1889 the head of the Department of Agriculture was given cabinet rank—a surprisingly late date for a farming nation, but official recognition, anyhow, of the importance of agricultural development. Between 1860 and 1900 over 400 million acres of virgin land came under the plow, and new agricultural machinery—the McCormick reaper, the Pitts mechanical thresher, and the Marsh harvester were all in use by 1858—was on the way to turning the farms into factories. With the mechanization of the land came a new rationalization of methods. Farming was less and less to be left to the traditional rule of thumb.

3

THE WINDS OF BIOLOGY

Walter Howard suggests that Luther Burbank can best be compared to the English horticulturist Thomas Andrew Knight (1759–1838). The peach and the grape could be produced in varieties that would grow well in England, Knight believed, just as the pear had been adapted to the English climate and the crab apple successfully introduced into Siberia. "Seedling plants . . . of every cultivated species sport in endless variety," he observed. "By selection from these, therefore, we can only hope for success in our pursuit of new and improved varieties."[1] In addition to peaches and grapes, Knight hybridized currants, apples, and pears to produce superior types. Crossing garden peas in the 1780s, he recorded the phenomenon of color dominance, perhaps for the first time, almost three-quarters of a century before Mendel.[2]

Mendel's discoveries were also anticipated by Knight's French contemporary Augustin Sageret (1763–1851), who aligned the traits of two varieties of the melon *Cucumis melo* in contrasting pairs and studied their inheritance. "Sageret not only confirmed the phenomenon of dominance and discovered the independent segregation of different characters," writes Ernst Mayr, "he was also fully aware of the importance of recombination."[3] Like Knight, however, Sageret failed "to ask questions about underlying mechanisms" and to develop a genetic theory. The reason, Mayr surmises, was that these experimenters thought in terms of *species* rather than variable *populations*, and inheritance can only be grasped in populational terms.

The traditional view of species derived from the Old Testament, but it acquired weight from perceptions—possibly inaccurate—of what the Greek philosophers Plato and Aristotle had taught. Plato offered a theory of "forms," or ideal essences, that exist independently of the world of appearances and are the patterns used by the demiurge (God) in creation. The goal of the created may be seen as the realization of this ideal nature, rather than development and transformation. This is clearly no basis for a theory of evolution in the modern sense.

Aristotle, "the first great naturalist of whom we know," seems to have pictured the cosmos as a static hierarchy of unchanging categories, in which there is neither creation nor extinction and "man and the genera of animals and plants are eternal."[4] In questioning Plato's theory of forms, Aristotle laid the foundations of scientific reductionism, but there was no room for development in his scheme of things either, and for two thousand years his immense authority blocked the emergence of a theory of evolution.

In the eighteenth century, however, the emerging science of geology dealt a blow to the Aristotelian world view by producing an awareness of prehistoric species that had differed vastly from existing ones. At first it was generally assumed that these had been obliterated by the biblical Flood, or similar global catastrophes, and that new species had been created outright to replace them. Few doubted that creation, or at least the latest round of it, had taken place in about the year 4004 B.C., much as described in the Book of Genesis. In opposition to this, however, geologists pointed out that on the available evidence, the earth was immensely older—perhaps even hundreds or thousands of millions of years old—and that worldwide disasters seemed improbable. The time was ripe for a new synthesis.

It came, appropriately enough, from revolutionary France, where Jean Baptiste Pierre Antoine de Monet, Chevalier de Lamarck (1744–1829), was professor of invertebrate zoology at the newly organized Muséum d'histoire naturelle in Paris, an institution that had evolved out of the medieval Jardin des simples, or garden of medicinal plants. In place of catastrophism-creationism,

Lamarck proposed that living things had gone through a process of gradual evolution over the ages, taking the form of a "branched series" (*série rameuse*). Classes, orders, genera, and species, he thought, were more in the nature of intellectual conveniences than realities. "Never forget that nature recognizes none of them," he told his students.[5] Lamarck argued that new needs (*besoins*) imposed by the environment brought about the development of new organs and traits. New species evolved when these were in due course transmitted to the individual's offspring. It has been suggested, too, that Lamarck possessed "a very clear idea of Natural Selection."[6]

At that time, no one would have thought to doubt that acquired characteristics could be inherited. It was simply taken for granted. The radical aspect of Lamarck's thesis was the idea that the process took place in conformity to changing environmental needs and gave rise to hierarchical evolution. It is very difficult for us today to grasp how novel this suggestion was. In place of Aristotelian stasis, Lamarck offered a view of life as a dynamic sequence developing over immense periods of time. It was this that made him "the creator of the concept of biology."[7]

Paradoxically, it was because of his belief in the inheritance of acquired characteristics—the least original aspect of his thought—that Lamarck was damned by modern science. In fact, it was not at all for this reason that he originally suffered eclipse. Rather, as "a man of the Revolution," he fell foul of the changing political climate: into disfavor under Napoleon and disregard when the Bourbon monarchy was restored. Nonetheless, his ideas found their way across the Channel, where they were introduced to the British public by the geologist Charles Lyell, who has been called perhaps "the greatest single influence in the life of Charles Darwin."[8]

When Lamarck died in 1829, impoverished and long neglected, it fell to the great comparative anatomist and paleontologist Georges Cuvier (1769–1832), "one of the most catastrophic of catastrophists,"[9] to deliver the customary eulogy in the French Academy. Cuvier managed to distort his rival's views so thoroughly as to give the impression that Lamarck had believed that animals somehow transformed themselves because they *wanted* to do so. This absurd idea was let loose in the English-speaking world when Cuvier's libelous valediction was translated and published in the *Edinburgh New Philosophical Journal* in 1836. It was lent added force by the misrendering of the French word *besoin* as *want* rather than *need.* The general impression was conveyed to a new generation of biologists—Darwin among them—that Lamarck had simply been insane. In England Lamarckism was also rejected because of its author's supposed political coloration. "Evolution was suspect as French atheism," writes Loren Eisely. "English naturalists discussing the species question disavowed Lamarck with the ritual regularity of communists abjuring deviationist tendencies today."[10] In later generations, embattled and ideologically committed Darwinists joined the assault for quite other reasons.

Lamarck's fate is a striking example both of the workings of the irrational in human affairs and of the remarkable gulf that separates the mind of England from the mind of France. Ironically, there was room for both Lamarck's evolutionism and Cuvier's catastrophism. Cuvier's theory of cataclysmic changes was, after all, based not on the Bible but on solid paleontological and geological evidence. We are now able to see that, in their different ways, both men were right about a great deal.

In 1819, William Herbert, the first English-speaking investigator to become aware of the work of Josef Kölreuter, also challenged the concept of the species as a fixed and immutable category. It was then believed that fertile offspring proved that the parents were of the same species, and conversely that sterile progeny showed that they were different. This rule of thumb was unreliable, Herbert argued. In his view, "fertility depended much upon circumstances of climate, soil and situation. . . . more upon the constitutional than the closer botanical affinities of the parents." In other words, analogous environmental adaptation, rather than formal similarity, was the important factor in successful interbreeding of botanically related types to produce a new line. "Any discrimination between species and permanent varieties is artificial, capricious, and insignificant," Herbert said.[11] But he offered no general theory.*

* Herbert defined the role of the horticulturist in a peroration that would have delighted Luther Burbank:

> To the cultivators of ornamental plants, the facility of raising hybrid varieties affords an endless source of interest and amusement. He sees in the several species of each genus that he possesses, the materials with which he must work, and he considers in what

The debate over evolution was not publicly taken up again until 1859, when one of the most revolutionary works of all time appeared in the London bookshops: Charles Darwin's *The Origin of Species*. Darwin had been developing his theory for many years. It had germinated, he tells us, in 1838, when, "I happened to read for amusement Malthus on *Population*, and being well prepared to appreciate the struggle for existence which everywhere goes on from long-continued observation of the habits of animals and plants, it at once struck me that under these circumstances favourable variations would tend to be preserved, and unfavourable ones to be destroyed. The result of this would be the formation of new species."[12]

How much Darwin may have owed to Lamarck and other precursors may be endlessly speculated. No one works in a vacuum, and the greatest thinkers tend to draw on the widest range of sources. A few of Darwin's may be noted. "Perhaps no other botanist influenced Darwin's thinking more than William Herbert," Mayr suggests.[13] And Loren Eiseley has argued that many of the main concepts from which Darwin built up his thesis—the struggle for existence, variation, natural selection, and sexual selection—were to be found in papers published by the naturalist Edward Blyth in 1835–37. Blyth and Darwin were friends, and Darwin had almost certainly read Blyth's essays. The crucial distinction between the two men was that Blyth was a special creationist and Darwin was an evolutionist. Blyth regarded natural selection as a conservative mechanism by which the stability of nature was maintained. Darwin saw it as the dynamic of change. Natural selection, Darwinism contended, could give rise to all the improbabilities to be observed in nature. There was thus no need

> manner he can blend them to the best advantage, looking to the several gifts in which each excels, whether of hardiness to endure our seasons, or brilliancy in its colors, of delicacy in its markings, of fragrance, or stature, or profusion of blossom, and he may anticipate with tolerable accuracy the probable aspect of the intermediate plant *which he is permitted to create*; for that term may be figuratively applied to the introduction into the world of a natural form which has probably never before existed in it. [Emphasis added.]

Since it came from a clergyman (Herbert rose to be Anglican dean of Manchester), Burbank might have found some comfort in this defense against those who were to accuse him of blasphemy for speaking of his "new creations."

to invoke design or divine intervention to account for the wonders of life. It was this aspect of Darwinism that caused such an uproar among the theologians.

Darwin, a cautious man, delayed publishing his theory for many years, while gradually assembling more and more data. It was not until 1856 that, at the urging of Sir Charles Lyell, his long-time mentor, he even began to write out his ideas in full. Early in the summer of 1858, however, he received an essay entitled "On the Tendency of Varieties to Depart Indefinitely from the Original Type," containing an almost identical hypothesis, from Alfred Russel Wallace, who was at the time travelling in the Malay archipelago. Wallace knew that Darwin had long been at work on the subject and had hastened to notify him of his breakthrough.

Darwin was stunned by the coincidence, but his priority was established by a letter he had written to Asa Gray, the leading American botanist, the previous year. This was read to the Linnean Society along with Wallace's paper on July 1, 1858, and publication in the society's *Proceedings* followed. Darwin pushed ahead with his work, and after "thirteen months and ten days' hard labour," *The Origin of Species* was finished. His book was to establish Darwin as the best-known natural historian of all time. Darwinist biology was ultimately to become a pillar in the great structure of scientific determinism, matching the "hard" sciences, physics and chemistry. But in 1858 the theory of evolution by natural selection had a long haul ahead of it. Indeed, it was not until the 1930s that natural selection was generally accepted among biologists as the evolutionary mechanism. "The early Mendelians. . . . thought of Mendelism as having dealt a death blow to selection theory," notes R. A. Fisher. "Up to the 1920s and 1930s, virtually all the major books on evolution . . . were more or less strongly anti-Darwinian," comments Mayr, himself one of the articulators of the "evolutionary synthesis" that reversed the trend.[14]

The eclipse of selection theory in the three decades following the rediscovery of Mendel's work in 1900 is largely forgotten today. In popular mythology, the rise of Darwinist biology is seen as an uninterrupted progress from strength to strength after the routing of Bishop Wilberforce and the forces of ignorance by the great man's janissaries. The reason for this misconception is probably that notwithstanding the decline of interest in selection theory among the community of biologists, two entirely unscientific versions of Darwinism continued to thrive. Darwinism acquired a currency in the political marketplace that it temporarily ceased to have in the laboratory.

In the first place, evolution by natural selection became the standard under which generations of ardent atheists made war on organized religion and on metaphysics in general. Darwin's importance to the evolving creed that was dubbed "scientific materialism" may, for example, be gauged from the fact that Karl Marx actually sought to dedicate the English edition of *Das Kapital* to him. The honor was declined—having verged on becoming a clergyman, Darwin did not wish to be associated with a book that attacked Christianity—but he entered the Marxist pantheon in spite of himself.

Another successful late-nineteenth-century ideology was the so-called "Social Darwinism" propounded by the utilitarian theorist Herbert Spencer, which sought to apply the concept of natural selection to human society. At the urging of Alfred Russel Wallace, Darwin had adopted Spencer's phrase "survival of the fittest" as being easier to understand than "natural selection." This was a mistake. As Sir Gavin de Beer has pointed out, "the 'fittest' can survive without any evolution taking place at all."[15] H. G. Cannon more bluntly asserts that "survival of the fittest" is simply a tautology.[16] It is a truism that contributes nothing to our understanding of the evolutionary process.

Spencer's "Social Darwinism" complied wonderfully with the ideological needs of the conservative business barons who ran the United States, however, and it enjoyed a tremendous vogue toward the end of the nineteenth century. Thus John D. Rockefeller justified the Standard Oil trust to his Sunday school class as "merely a survival of the fittest. . . . The American Beauty rose can be produced in the splendor and fragrance which bring cheer to its beholder only by sacrificing the early buds which grow up around it. This is not an evil tendency in business. It is merely the working-out of a law of nature and a law of God."[17]

It was doubtless no accident that Rockefeller's analogy was a horticultural one. Luther Burbank was at the time America's most highly visible Darwinian. His ghost-writers and promoters constantly harped on the fact that Darwin was the master in whose steps Burbank walked. "Read Darwin first, and gain a full comprehension of the meaning of Natural Selection," one of them has him say. "Then read the modern Mendelists in detail. But then—go back to Darwin."[18] We may be almost certain that Burbank was not responsible for this portentous utterance in its printed form, but the distinction between Darwin and the "modern Mendelists" may well be his. He realized, as many did not, that there was a great gulf between Darwinism and neo-Mendelism. And his own views no more coincided with contemporary biological dogma than they did with simplistic Social Darwinism of Rockefeller's kind.

Burbank probably never read *The Origin of Species*—he admitted in a letter written in January 1909 that he had not yet done so. He had learnt his Darwinism from a subsequent book of the master's, *The Variation of Animals and Plants under Domestication*, which reached him in Lancaster, Massachusetts, in 1868, the year of its original publication. He was nineteen when he read it, and his enthusiasm was unbounded. "It opened a new world to me," he says. "While I had been struggling along with my experiments, blundering on half-truths and truths, the great master had been reasoning out causes and effects for me and setting them down in orderly fashion, easy to understand, and having an immediate bearing on my work! I doubt if it is possible to make any one realize what this book meant to me."[19]

Later Burbank also acquired a copy of Darwin's *The Effects of Cross and Self Fertilization in the Vegetable Kingdom*, published in 1876. By that time, however, he had launched his career and was living in California. Burbank's Darwin was to all intents and purposes the Darwin of the *Variation*, a book based on the great mass of material Darwin had accumulated prior to writing the *Origin*. In it, Darwin restated his principle of natural selection and argued against the possibility of design in variation. He also attempted to come to terms with the most serious criticism that had been leveled against his system, that advanced by the engineer Fleeming Jenkin. In the *North British Review* of June 1867, Jenkin had demonstrated that the "blending" view of heredity—and, for that matter, the theoretical assumptions of animal breeders up to that time—was necessarily invalid. Single variations would not be perpetuated but swamped by the mass of population. Darwin was temporarily stumped. In 1869 we find him writing pessimistically to Wallace, "Jenkin argued . . . against single variations ever being perpetuated, and has convinced me."[20]

Though Darwin seems to have envisaged the characteristics of progeny as something more than a mere blend of parental attributes, this was the accepted view at the time, and he was in possession of no clear alternative. Mendel's monograph establishing that inheritance is not a fusion of parental characteristics, but that traits are inherited independently and preserve their independent identities in succeeding generations, was unknown to Darwin. This was bad luck, since it had been published in 1865, several years before the *Variation*, and copies undoubtedly reached England. The statistical approach to the study of hybridization that Mendel adopted was too much of a novelty to find acceptance with contemporary biologists, however, and his insight long went unheeded.

In the *Variation*, Darwin opted for a theory of particulate inheritance he called "pangenesis." This postulated that particles he dubbed "gemmules" were thrown off by the cells of the body and found their way into the reproductive system, where they formed the hereditary constitutents of sperm and ovum.[21] Environmental conditions affecting the parents are thus able to modify inheritance: acquired characteristics may be preserved in future generations.

Faced with Jenkin's argument against single variations ever being perpetuated under the "blending" theory, Darwin "panicked and ran straight into the opposite camp," says C. D. Darlington. "Lamarck became a posthumous Darwinian."[22] But H. G. Cannon points out that Darwin had expressed distinctly Lamarckian opinions in *The Origin of Species* itself, commencing with the first edition. In the sixth edition, Darwin actually inserted a discussion of the evolution of the giraffe in terms of natural selection "combined no doubt in a most important manner with the inherited effects of the increased use of parts."[23] (The lengthening of the giraffe's neck by stretching is, of course, the classic chestnut offered by countless schoolteachers to discredit Lamarck.) "Without doubt Darwin was a believer in the inheritance of acquired characters, that is, according to our contemporary idea, a 'Lamarckian' at the time of the first publication of the *Origin*," Cannon says.[24]

In Darwin's day, little was known about the physical mechanisms of inheritance. The idea that acquired characteristics were passed on to progeny was ancient and possessed no small biblical authority. ("Adam's sin is the unforgettable archetype for the transmission of an acquired character," Madeleine Barthélmy-Madaule observes.)[25] The question was *how* traits were passed on, and some sort of particulate inheritance seemed logical. Ten years before Mendel's long-forgotten papers came to his attention, Hugo de Vries published his own theory of "intracellular pangenesis," in which he proposed that hereditary particles corresponding to the different adult characteristics must exist in all the cells of the organism—a remarkable foreshadowing of the modern concept of the gene. "These factors are the units which the science of heredity has to investigate," he wrote. "Just as physics and chemistry are based on molecules and atoms, even so the biological sciences must penetrate to these units in order to explain by their combinations the phenomena of the living world."[26]

In the late nineteenth century, the new science of cytology—the branch of biology that deals with cellular structure, evolution, and function—underwent astonishingly rapid development. In the space of less than twenty-five years, the whole basis of genetics was brought into being. In 1888 Wilhelm Waldeyer introduced the term *chromosome* to describe the threadlike bodies in the cell nucleus, visible in stained material, that in dividing initiate the formation of new cells. In 1900 Mendel's work was simultaneously rediscovered by De Vries, Carl Correns, and Erich Tschermak. The term *genetics* was first used by the English biologist William Bateson in 1906, and in 1909 the key words *gene*, *phenotype*, and *genotype* were introduced by the Danish botanist Wilhelm Johannsen.

In the mid-1880s the German zoologist August Weismann had begun promulgating a remarkable doctrine. Weismann had come to the conclusion that the hereditary substance that resided in the chromosomes, which he called the germ plasm, was entirely independent of the rest of the body. The germ plasm is not recreated in each individual, he asserted, but is passed on as a continuous heritage. Acquired characteristics cannot be inherited in this view, "and we must entirely abandon the principle by which alone Lamarck sought to explain the transformation of species—a principle of which the application has been greatly restricted by Darwin in the discovery of natural selection, but which was still to a large extent retained by him."[27]

Proposed in the infancy of cytological studies, this idea of an almost metaphysical discontinuity between the germ cells and other somatic cells was daring to say the least.* Nonetheless, it became generally accepted among biologists. Perhaps the reason for its elevation to the status of an axiom was that it so perfectly suited the prevailing climate of reductionism. It eliminated the messy notion of an indeterminable feedback between genotype and phenotype. Instead, dovetailing neatly with Mendelism, it gave authority to that "set of demons, the genes, capable of extricating the theorist from any difficult situation."[28] Only today is the Weismann doctrine beginning to be systematically reassessed.[29]

In the early 1900s, battle raged among biologists over the question of whether mutations were the prime instruments of evolution or gradual accumulation of fluctuating variations within species was the decisive factor. When De Vries visited Burbank in the summer of 1904, he was already thinking of ways to influence the germ plasm experimentally using X-rays and radium. In his recent book *The Mutation Theory* (1901–3), he had outlined a new view of species as complexes of discontinuous, independent, and independently inheritable traits, subject to variations that took place by leaps and bounds, which he called "mutations," adding a new concept to the vocabulary of science.

Burbank does not seem to have read much of the scientific literature of the day, and he probably never used a microscope in his life. He based his work on his own experience in the field and on Darwin—the "Lamarckian" Darwin he knew from the *Variation*. He was not much impressed by De Vries's theory of massive mutations, and of periods of mutation in the life of species. He simply saw no need for it. "The varied tribes of evening primroses which Professor de Vries developed in his gardens at Amsterdam were overwhelmingly suggestive of various and sundry forms of hybrid plants that I myself have developed year after year in my experi-

* "It has been said by many modern writers that Weismann carried this inviolability principle too far, but it should be remarked in simple justice that since his works are no longer read in great detail, his own qualifications upon this point have been forgotten," observes Loren Eiseley. "He was willing to concede that the germ plasm was probably not totally isolable from influences penetrating it from the body, but that such influences 'must be extremely slight'" (*Darwin's Century: Evolution and the Men Who Discovered It* [Garden City, N.Y.: Doubleday Anchor, 1961], pp. 218–19). That is precisely the point: the qualifications were forgotten, and the doctrine became dogma.

mental gardens at Santa Rosa," he is quoted as saying. "Over and over again, hundreds of times in the aggregate, I have selected mutants among my plants, and have developed from them new fixed races. But in the vast majority of cases I knew precisely how and why these mutants originated. They were hybrids; and they were mutants *because* they were hybrids. And so from the outset I have believed that Professor de Vries' celebrated evening primroses had the same origin."[30]

And he was right. De Vries's unusual primroses, on which so much of his mutation theory rested, were mostly hybrids. Subsequent study has identified them as being of several different kinds. Some were a consequence of incomplete chromosome pairing, some were polyploids, involving changes in the number of chromosome sets; only two were mutants in the modern sense of the term. "The mutants of Oenothera are therefore nothing more than symptoms of its peculiar hybridity and as such are of little significance in evolution," observe Darlington and Mather.[31]

There was no reason, of course, for De Vries to concede to Burbank's opinion, which was not substantiated until years later.* But there is little doubt that he took Burbank very seriously indeed. Returning to San Francisco in 1904, he delivered a speech (in German) in which he referred to Burbank as an "innately high genius" (*eingeborenen hohen Genie*).[32]

What was it that made De Vries, who "dominated the thinking of biology from 1900 to 1910,"[33] say things like that about the gardener of Santa Rosa? All else aside, perhaps the great Dutch biologist already uncomfortably sensed the closing in of determinism in genetics. The applications of neo-Mendelism were to prove immensely fruitful in practice (in the work of Shull and Jones with hybrid corn, for example), but the Weismann doctrine of the unbroken continuity of the germ plasm has profoundly nihilistic implications. If the possibility of the inheritance of acquired characters is denied, human evolution must, biologically speaking, be at an end once the pressure of natural selection has been culturally lifted. The sterilities of the nature-nurture debate are what remain.

In this sense, De Vries may have seen in Burbank a symbol of

* Modern views of the *Oenothera* complex date from H. J. Muller's study of "balanced lethals" in his 1917 paper "An Oenothera-like Case in Drosophila." See also R. E. Cleland, *Oenothera: Cytogenetics and Evolution* (New York: Academic Press, 1972), on the subject.

hope. He may have *wanted* Burbank to be right. "It is difficult for man to believe that he has *no* genetic future, or if he does, a future without direction," observes E. J. Steele.[34] Burbank's genius was practical and intuitive, not theoretical. His conception of the way inheritance worked in no way resembled neo-Mendelian orthodoxy, but it seemed to be confirmed by a host of tangible achievements. If his experience ran contrary to the theories of the scientists, he did not let that faze him. He went his own way regardless of the prevailing winds of biology.

4

NEW ENGLAND ROCK

The past, it has been observed, is another country. The aphorist probably understated the case. The latter half of the nineteenth century, in so many ways the mold in which our own way of life was originally cast, is already infinitely more remote from us than the surface of the moon. And, unlike the moon, it recedes with every succeeding generation.

"Between 1850 and 1900 nearly every one's existence was exceptional," wrote Henry Adams. For Americans in particular the relative velocity of change was as great or perhaps even greater than it is today. In 1846 (the "year of decision" Bernard De Voto was to call it) war with Mexico had established the geography we now take for granted. The frontiers of the spirit, elusive, eternally debatable, and yet just as crucial to the developing nation, were less easily determined. Fifteen years later came the terrible half-decade of the Civil War. Luther Burbank was eleven years old at the outbreak of fighting. He was just sixteen in April 1865 when Lee surrendered to Grant at Appomattox Courthouse. By then he was already earning his own living. He had grown up while the world seemed to be falling down, and there was room for every hand in the labor-starved workshops of a nation divided against itself.

Looking backward from a distance of fifty years, Henry Adams wondered "whether, on the whole, the boy of 1854 stood nearer to the thought of 1904, or to that of the year 1."[1] In the essentials, he concluded in favor of the earlier date, so great were the revolutions he had seen, in science and in art, in daily life and in public life. And if this were true of him, grandson of one president, great-grandson of another, scion of the greatest American family,[2] graduate of Harvard College, and seasoned visitor of the capitals of Europe, how much more so of a boy born of plain people at Lancaster, Massachusetts—not that many miles from the Adams's ancestral home at Quincy, true, but a fair eternity from its Queen Anne mahogany paneling and Louis XVI furniture?

Luther Burbank was born at 11:57 P.M.[3] on March 7, 1849, three days after the inauguration of "Old Rough-and-Ready" Zachary Taylor, who had earned his hero's laurel and presidential candidacy across the Rio Grande in the Mexican War. The Gilded Age, as Mark Twain would call it, had not quite begun. Like President Taylor, who now pastured his war-horse Old Whitey on the White House lawn, the United States was a rough-and-ready nation, so busy growing that even its capital, begun over fifty years before, was only half built.

Taylor, the Whig candidate, had carried the Massachusetts vote. But he was also a slave-holding Virginian, and already the party had split into "Conscience Whigs" and "Cotton Whigs," foreshadowing the struggle that was to come. From the little we know of Luther's father, it seems safe to suppose that he was for "conscience" and, like most other antislavery Whigs, shortly became a Republican. Luther would be a Republican in his turn until 1916, when he came out for Woodrow Wilson.

The best and only full account of his family and youth is to be found in *The Early Life and Letters of Luther Burbank*,[4] a somewhat mawkish little book published by his sister, Emma Burbank Beeson, in 1926, the year of his death, when she herself was seventy-two. The Burbanks, Emma says, were of English extraction,

> although the name Bermbank (as it is sometimes spelled) glimpses Holland, and no doubt a branch of the family were, in the fifteenth century, Belgian-Dutch.
>
> We are told of five Burbank brothers coming to the New World from the North of England, but the first authentic record is to be

> found in the Custom House at Boston, Mass. Joseph Borebank came in the ship *Abigail* from London in 1635; and John Burbank, from whom is traced in direct line our family, was made a voter at Rowley, Mass., in 1640.
>
> The following brief record of our line in the Burbank genealogy shows the rather unusual maturity of parents, as five generations cover a period of over two hundred years; John Burbank a voter at Rowley, Massachusetts in 1640; his son, Caleb Burbank, was born 1646; when 35 years of age his son Eleazer Burbank was born 1681; when 26 years of age his son Daniel Burbank was born 1707; when 39 years of age his son Nathaniel Burbank was born 1746; when 49 years of age his son Samuel Walton Burbank was born 1795; when 54 years of age his son Luther Burbank was born 1849.

She lists the various spellings of the name in the roll call of soldiers in the Revolutionary War: Birbank, Burbanck, Burbankes, Burbanks, Burbect, and Burbank. The New England Burbanks, she says, were generally farmers, manufacturers, teachers, and clergymen, for "throughout their generations it has been easier for them to solve a mathematical problem than to drive a nail, and they could make a political argument more successfully than they could make a flower to flourish." Few New England families, she claims with pride, "were more eminently represented in the learned professions, in civil enactments, and public reforms."

Nathaniel Burbank, Luther Burbank's grandfather, born at Sutton, Massachusetts, in 1746, married a widow, Ruth Felch Foster, and fathered seven children. At the time of his marriage he was a paper maker at Harvard, but in 1797 he moved to Lancaster, not far off, where he bought the farm on which Luther was to be born. Until his death in 1818, he manufactured bricks and pottery there, making use of the fine clay on his property. He was, reports his granddaughter, "a quiet, peace-loving man." For all that, he had a mind of his own. When his wife and sons left the Unitarian church at Lancaster to join the more recently established Baptist church at Still River, a small village in the opposite direction from home, he continued to go on foot each Sunday to his own church. Asked why he did not go with his wife and sons, he replied that the first parish in Lancaster had always suited him, and he saw no reason to change. The will he left gives a vivid picture of rural life in those days:

In the name of GOD, Amen.

I, Nathaniel Burbank of Lancaster, being weak in body but of sound mind and memory, blessed be God for the same, do this twentieth day of February, in the year one thousand eight hundred and eighteen, make and publish this my last will and testament in manner and form following; that is to say:

I give to my wife Ruth Burbank the use of a good cow, to be kept summer and winter; and also eighty weight of beef yearly, and seventy weight of pork, and nine bushels of corn and six bushels of rye, one bushel and a half of wheat and all kinds of summer and winter sauce sufficient for her use, and also the use of one-third of my dwelling house during her natural life, the said third to be chosen by her; and also a sufficiency of good firewood cut fit for the fire at the door, a horse and carriage for her use, and also a good doctor procured for her when needed, and paid, and all her grain carried to mill and the meal returned to her, and also to be comfortably clothed; and also to have the use of all my furniture during her natural life, and at her decease the said furniture is to be equally divided between my two daughters, Mehitable Barrett and Lucy Ball.

Item: I give to my beloved son, Daniel Burbank, the use of my shoemaker's shop to work in whenever he pleases, and an equal right to work in my brickyard for his own emolument with my two sons hereafter named, and to pass and repass to and from the same, and also one acre of land where the old barn used to stand, beginning at the road leading to Harvard and bounding on land belonging to the sons of Timothy Lewis, deceased.

Item: I give to the heirs of my beloved son Caleb Burbank, deceased, the sum of fifty dollars, to be well and truly paid to them by my executor after my decease.

Item: I give to my beloved son Nathaniel Burbank the sum of fifty dollars, to be well and truly paid to him by my executors after my decease.

Item: I give to my two beloved sons, Samuel Walton Burbank and Aaron Burbank, all my lands and buildings and stock and farming tools not before disposed of in this my will whatsoever, whereof I shall die seized in possession of, by their well and truly paying all my just debts and funeral charges.

Item: I make and ordain them to be executors of this my last will and testament.

In witness whereof I, Nathaniel Burbank, have to this my last will and testament set my hand and seal the twentieth day of February, in the year of our Lord one thousand eight hundred and eighteen.

> Signed, sealed, published and delivered by the said Nathaniel Burbank as and for his last will and testament, in the presence of us who at his request and in his presence, and in the presence of each other, have subscribed our names as witnesses thereto.
> NATHANIEL BURBANK [Seal]
> CONSIDER STUDLEY
> JABES DAMON
> DARBY WILLARD

The language itself is modern and literate; but in every other respect (even to the names of the witnesses), it might be a document of the old Massachusetts Bay Colony. In Lancaster village, the nineteenth century had not yet properly commenced.

Luther's father, Samuel Walton Burbank, who, with his brother Aaron, inherited the land, buildings, stock, and farming tools at Lancaster, was born at Harvard in 1795 and was only two years old when the family moved to the farm. In 1818, shortly before Nathaniel Burbank's death, Samuel and Aaron built the Burbank home. A "large, square, brick house, divided by a hall through the center and with an ell kitchen on each side," it was paid for by the manufacture of brick used in the construction of a church in Lancaster two years before. "In this home," says Emma, "father spent the remaining fifty years of his life. The home is still occupied* and stands under the swaying branches of a great elm tree, which was set out by father in 1821, about the time he brought his bride, Hannah Ball of Townsend, Mass., to the home."

Hannah died after bearing him eight children, and he presently remarried. His second wife, Mary Ann Rugg, had two children by him, but neither she nor they long survived, and in June 1845 he married Olive Ross of Sterling, Massachusetts, who would bear him an additional five. The first two of Olive's children died in infancy. Luther, his father's thirteenth child and the first of Olive's to survive, was born in the fourth year of their marriage. Another son, Alfred, and a daughter, Emma, Samuel Burbank's fifteenth

* In *Luther Burbank: The Wizard and the Man* (New York: Meredith Press, 1967), Ken and Pat Kraft write: "The land it stood on is now part of the Army's Fort Devens, and the house was razed during the warm-up for World War II in 1941, to make way for a firing range" (p. 17).

and last child and the author of *The Early Life and Letters of Luther Burbank*, were born later.[5]

The terrible mortality rate in the Burbank family—only half of the children surviving their teens—was nothing uncommon in those days; but it may go some way to explaining Luther's lifelong tendency to hypochondria. The graveyard where seven of his father's children lay buried was never very far away in his youth, and the anniversaries of their deaths came round all too often. Though they had died before he was born, the impression of mortality must have been powerful in a sensitive child. Fatal illness was only too familiar, and any illness might prove fatal. The century itself, with its thriving horse doctors and host of patent medicines, was more than a little hypochondriac, with excellent reason.

But the Burbanks seem to have been fortunate as families went. "My father," Luther says in his autobiography, "was a New Englander of pure and unmixed physical strain, but it was not the shallow soil, the rocky structure below, the hardness of labor, the rigor of the winters, and the austerity of the people about him that he saw; he was a man of imagination and a facile mind, and he loved beauty and the sunshine and pleasantness of the land, in its garments of spring and autumn, and was influenced by them."[6] Emma adds to this that Samuel Burbank "loved his children tenderly, and gave to each the best care and education in his power. His love of reading and classic taste kept the home supplied with good literature. The children had books and papers suited to their taste and ability."

What were the books they had? Emma does not tell us. There must have been the Bible. Almost certainly there was a copy of Noah Webster's "Blue-Backed Speller," over thirty million of which had been sold by 1859. Very likely there were the works of Thoreau, Emerson, Whittier, Lowell, Hawthorne, and Cooper. There was probably *Uncle Tom's Cabin*; perhaps *Hiawatha*, *The Age of Fable*, and Bartlett's *Familiar Quotations*, all published in 1855. Sir Walter Scott, the enormously popular Washington Irving, and even Susan Warner's sensationally successful tearjerker *The Wide, Wide World*, with more than half a million copies in print, may have been there. Of magazines there may have been *Harper's*, launched in 1850, the New York *Knickerbocker*, in which the great New England writers frequently appeared, the Boston *Youth's*

Companion, and *Godey's Lady's Book*, in which the formidable Sarah Josepha Hale gave the world "Mary Had a Little Lamb."[7] *Godey's* was the country's best-selling magazine in 1850, and according to Emma her mother had gone to school with the Mary Sawyer who was the original heroine of that poem (though "Mary was her senior by several years, and the incident of the lamb at school which gave rise to the familiar poem occurred before her school life began"). The United States was in the midst of a media explosion, and by the middle of the century had already surpassed England in book and magazine sales. Massachusetts was the most literate of American states, and the Burbanks lived in close proximity to some of the most famous writers of the day.

Samuel Burbank, Emma says, was never arbitrary, and perhaps too kind and indulgent a father. "Someone once asked brother Alfred if father gave him a whipping for some misconduct. Alfred seemed loath to reply, for he knew he deserved it, but he at last said, 'No, but he offered to,' and that is as near as any of us ever came to a whipping." This indulgence he seems to have passed on to his son. All his life, Luther would have a soft spot for children.

Olive, Luther's mother, was the second child of Peter Ross and Polly Kendall Burpee, born on April 7, 1813. Her father was of Scots descent and a cabinetmaker by trade. But, Emma observes, he "was by nature a horticulturist, and was the originator of several new grapes. He was best known at the time by the superior fruits and vegetables produced on his place. He was a rather frail man, suffering all his life from attacks of heart disease, yet he lived to the advanced age of 87 years, retaining his ambition and enthusiasm to the last. He was clearing the brush from a very steep rocky hillside in order to set it with grapevines when he was called by death. He was found one bright summer day sitting on a big rock, apparently resting, as so often he had done, but life had quietly gone out. The active, executive wife and loving mother preceded him but a few years to the other life." Polly Burpee was of French descent (the name is a corruption of *beau pré:* beautiful meadow), and Emma remarks on the fact that she belonged to the family that included W. Atlee Burpee, the famous Philadelphia seedsman, who would one day be one of Luther's customers and champions.

In his autobiography, Burbank himself says of his mother that she was "shrewd and practical, of a nervous temperament, quick and impulsive, yet kindly, intelligent, and with a great love of her garden. She had an unusual bent for making things grow, whether from domesticated seeds or from bulbs or cuttings or roots found in the woods all about; one of my earliest recollections is of the beauty and peace and fragrance of that old-fashioned garden of hers." It was evidently from Olive, rather than from his father, that he had his love of gardening and feeling for growing things.

She was a woman of strong character. "Mother [it is Emma speaking] was of a very intense nature, walking with firm step and speaking with the falling accent." When only thirteen, she signed a temperance pledge and remained faithful to it throughout her long life. If this seems extraordinary, it should be remembered that in the New England of Olive Burbank's youth, rum, beer, and hard cider flowed uncommonly freely, as did a variety of brandies and "bounces" manufactured from cherries, peaches, and pears. Strong drink was not infrequently given to small children. "In accord with the general custom," Emma notes, "a large mug of 'hot toddy' was prepared on 'election day' and other holidays. Each member of the family, to the youngest child, was expected at least to sip from the mug. Olive at an early age became disinclined to follow the custom and refused to taste the liquor." She successfully inculcated at least one of her children with the temperance spirit: Luther would scarcely touch as much as a drop of alcohol all his days.

Emma gives a picture of life in the Ross home that she obviously had from her mother. It seems worth quoting as a portrait of a bygone age in which

> there was much of family life and companionship. When other tasks were remitted the father taught the boys to weave baskets from slender willow branches gathered from the banks of meadow streams; the girls braided hats from [the] finest of the wheat straws, which had been carefully selected and saved for this purpose. These hats were sold at the country store. In the attic chamber, where were kept the willow strands for basket-weaving and the unfinished work, heaped on the floor, in one corner, were glistening brown chestnuts; in another, big, rugged butternuts, and the plump hickory nuts, in larger quantity, filled a bin by themselves.
>
> Work and pleasure were combined in the gathering of the nuts,

> picking the apples and husking the corn, and also in the excursions for wild strawberries, low-running blackberries on the hillside, and huckleberries in the pasture lands. Olive was especially fond of these out-of-door tasks and of rambles in the woods gathering wild flowers, lupines, goldenrods and asters.
>
> Books then were few and even those not of a nature pleasing to childhood. Only homemade candles by which to read, yet around the glowing fireside, with popping corn, roasting chestnuts, and apples, there was enjoyed much more of neighborly hospitality than now exists.

It is a romantic picture that leaves out the harshness of life in the cruel New England winters, the perils of the inevitable sickbed, the stern morality of a Puritan society where, as Van Wyck Brooks put it, the congregations "followed the web of the sermons with a keen and anxious watchfulness, eager to learn the terms of their damnation." But it would be familiar enough to the average American of the 1820s, in whose country the farmer still outnumbered the townsman more than ten to one.

The town of Lancaster, incorporated in 1658, lies in Worcester County, Massachusetts. Here, Emma observes, "less than twenty-five years after the landing of the Pilgrims, a tract of land ten miles long and eight miles wide was purchased of the Indian chief Sholan of the Nashaway or Nashawogs." The Burbank homestead was about three miles north of Lancaster Center, just off the Harvard Road. It was "a typical New England home—the large, square brick house, standing well back from the street, beneath the swaying branches of a stately elm tree. Over the door climbed the white jasmine, while lilacs, roses, lilies, and other old-fashioned flowers filled the yard and looked from the windows of the cheery living-room."

Here lived the families of the two brothers, Aaron and Samuel Burbank. In addition to the house itself, their property consisted of some hundred acres of farmland, divided piecemeal between them in order that each might have the same share of the orchards, fields, pastures, and woodland. The brothers also maintained separate brickyards, which provided them with a supplementary source of income in the summer months. Aaron, a Baptist minister in addition to being a farmer and brick manufacturer, spent a good deal of his time tending to his congregation. Samuel devoted himself to the more mundane pursuits of farm, brickyard, and woodlot. Home life was strictly ordered. Each day's work began with family prayers and a text from the Bible. But, this being New England, there was also a lively interest in the secular thinking of the day. Worcester County lay at the heart of the Commonwealth of Massachusetts: it was no intellectual backwater.

"The home was filled with intellectual activity, being near Concord, then the center of American literature, famous as the home of Ralph Waldo Emerson and other men and women whose thought was influencing the world," Emma says. "All the household was greatly interested in the great leaders of the times—Lincoln, Webster, Sumner, Agassiz, the Beechers, as well as Emerson, Thoreau, Longfellow and the Alcotts, some of whom were personal acquaintainces and sometimes guests at the home." Luther himself told Edward Wickson that "my father was a great admirer and personal acquaintance of Ralph Waldo Emerson, Daniel Webster, Chas. Sumner and Henry Ward Beecher."[8] Beecher was a professional lecturer and pulpit orator, Webster and Sumner were Massachusetts politicians, and to supplement his income, Emerson lectured widely in Boston, Concord, and neighboring towns, of which Lancaster was one. There would have been nothing out of the ordinary about these men knowing Samuel Burbank, a prominent local citizen. The fact that Luther did not stress these friendships in his published writings makes it unlikely that they were more than casual acquaintances of his father's.

A closer, more constant presence was that of Aaron Burbank's oldest son, Levi Sumner Burbank, one of the first members of the American Association for the Advancement of Science. Levi Burbank was friendly with the great naturalist Louis Agassiz (1807–73) and had served as curator of geology at the Boston Society of Natural History. He was the author of several papers on geology and paleontology. When the Civil War broke out, he was president of Paducah College in Kentucky. Resigning this position for obvious reasons, he became principal of the Lancaster Academy in 1861, when Luther was twelve years old. He possessed a large geological collection, and he often took his young cousin with him on his excursions in search of natural history objects.

Levi lived at the Burbank homestead, and he was undoubtedly

instrumental in turning Luther's mind in the direction of natural history. He had a talent for explaining things in such a way as to make them "alive and exciting and never dull and tedious. . . . and had read more books and understood them better than any one else I knew." The boy was already at home with nature, but it was Levi, he afterward said, "who may have crystallized my formless thinking and shaped my vague theories until he had made me want to know, not second-hand, but first-hand, from Nature herself, what the rules of this exciting game of Life were."[9]

See, then, in the dark years of the Civil War, Levi and Luther roaming the countryside around Lancaster—a wooded landscape almost English in its "low, sloping hills, rich intervals, sparkling streams of pure water and enchanting little lakes like bits of blue sky dropped down to earth," Emma observes. She was the little girl who welcomed them back from their walks and, in listening excitedly to the story of their adventures, developed an admiration for her brother that lasted all her life.

When he was five, Luther started school in the little red-brick schoolhouse among the pine trees on what they called the Dingle Road, a crossroad connecting Shirley Road and Harvard Street. It was about half a mile from the Burbank homestead, and Luther's half-brother Herbert, who at twenty was the teacher, pulled him to school on his little handsled. His next teacher was his half-sister Jane (Eliza Jenny) who, Emma says, "had recently returned from Pennsylvania where she began teaching when only fourteen years of age." Studious though he was, even Herbert and Jane found it difficult to get Luther to recite his lessons in the schoolroom because of his extreme shyness. "Luther has his days of moroseness—but he is engenious [sic] and persevering yet lacks the generosity which would enable him to enjoy society," Herbert wrote to brother George in California in 1856.

Slightly built and not physically strong, Luther preferred to play with the girls, who often defended him against the bullying of bigger boys. But as he grew up, he showed a natural measure of mischief. He loved to skate and one year, when he was about eleven or twelve, dammed a trout brook that flowed through his father's fields on the pretext of increasing the cranberry crop on the peat bogs, but actually to make a skating pond. The scheme was successful, and a Christmas skating party was held on his homemade rink the following winter. Almost half a century later, Luther listed some of the ingredients of an ideal childhood, speaking from memory of his own: "Every child should have mud pies, grasshoppers, water-bugs, tadpoles, frogs, mud-turtles, elderberries, wild strawberries, acorns, chestnuts, trees to climb, brooks to wade in, water-lilies, woodchucks, bats, bees, butterflies, various animals to pet, hay-fields, pine-cones, rocks to roll, sand, snakes, huckleberries and hornets; and any child who has been deprived of these has been deprived of the best part of his education."[10]

When his twenty-five-year-old half-brother George grafted the apple trees in the orchard, the five-year-old Luther watched with intense interest. He could not understand why so many healthy branches had to be cut off, and George explained the process to him. He took great pleasure in watching the development of the grafts and spent a good deal of time in the apple orchard watching the buds grow, the pink and white blossoms unfurl to attract the bees, and the formation of the tiny apples. He was the one to carry back the news of the first faint coloring of the Early Junes and Williams. He soon learned to identify these and the other kinds—Baldwins, Greenings, and Russets—in the orchard. One of his early treasures was a book with descriptions and pictures in outline of different varieties of fruit.

There was one particular incident that had the effect of profoundly stimulating his inquisitiveness about the processes of nature. He was walking along in the woods, kicking at the snow and wishing that the winter were over. Suddenly in the white landscape he saw a green place in front of him. He could scarcely believe his eyes. The grasses grew tall and the shrubs and vines were fresh and springlike. Even the sunshine seemed to him "almost tropical in its white intensity." He rushed forward to find out if this miraculous oasis could be real and discovered a warm spring that carried enough heat to the surface to keep the ground roundabout free of snow and give the vegetable life a foothold even in the middle of winter.

> I studied the phenomenon intently. Very dimly there began to grow in my mind vague questions as to how these plants and grasses and vines, their neighbors and cousins and brothers all dead and withered under the snow, or else dormant and waiting for spring, could adjust

> themselves to this summer-in-winter environment. Why, I asked, didn't they follow the traditions of their families and die with the fall, or droop and shed their leaves and hibernate, no matter what the warm water brought to them in the way of a miracle of equable temperature? Certainly the lycopodiums, the beautiful trailing partridge berry, the sedges and grasses, and here and there an early sprouting buttercup, should have known better, the way they had been raised and with their decent seven-months-of-summer ancestry behind them, than to flaunt themselves so shamelessly in this unfilial winter blooming![11]

To this incident Burbank later attributed his first notion of the power of environment in plant variation. When he got home that evening, it was late, and his family were not much impressed by the miracle he had found. But to him it was the most wonderful thing that had happened in his young life. Levi Burbank discovered a disciple eager to learn what he had to teach when he came back from Kentucky not long afterward with his collection of geological specimens and his fondness for botanical expeditions. Natural history became Luther's passion.* In January 1864, we find him writing eagerly to his half-brother George in California: "We all want to see you very much. When you come will you bring seeds of some of the plants that do not grow here, and some of the minerals, too?"

Physical labor was early on a regular part of Luther's life. "I used to team bricks with an ox team with a horse in front to the Clinton Wire Mills and the Clinton Gingham Mills and to many other manufactories in the towns around Lancaster before I was fifteen years of age," he later recalled. In the fall of 1864, however, he entered Lancaster Academy as a student.[12] For the next four years, he spent the winter months studying and worked in the summers at the Ames Plow Works in Worcester, where his uncle Luther Ross was a foreman, learning the patternmaker's trade, for

* Frederick W. Clampett says that Louis Agassiz was a close personal friend of Luther's father and that in 1863, "when Luther had attained his fourteenth year and Agassiz was just fifty-six years old, he formed his acquaintance, so that for ten precious years he sat at the feet of that great master" (*Luther Burbank, "Our Beloved Infidel"* [New York: Macmillan, 1926], p. 19). This seems improbable, and Agassiz was in any case a resolute anti-Darwinian. Clampett also says that Burbank recalled "with remarkable distinctness" many long conversations he had had with Emerson when a child (ibid.).

which he showed considerable aptitude. There was a bonus in that his uncle had a garden and even apparently did a little horticultural experimenting as a hobby (his father, Peter Ross, it will be remembered, was known as the originator of several new varieties of grape). On occasion, the young Burbank was allowed out of the workshop to labor among the seedling grapes and the rhubarb.

He remained a quiet, reticent boy, particularly timid with strangers. Once when his mother was visiting him at his uncle's house and they sat talking in the living room, his aunt, who was preparing supper in the kitchen, pushed open the door a little so that she "might hear the sound of Luther's voice," as she explained when detected. She had heard him speak so few words since he had come into the household.

Luther did not stay on at the Lancaster Academy for a fifth year, but one wonders how much of its elaborate syllabus he was able to cover studying only part time. The emphasis appears to have been heavily on classics, and there is no evidence that he knew much Greek or Latin. Emma asserts that he did well in mathematics and English composition but found the afternoons, when the students were required to declaim, a torment. "After one or two ineffectual attempts to stand and deliver a perfectly committed address," she says, "he was excused from this exercise with the provision that he write double the required number of themes, and thenceforth life at the academy was a pleasure to him."

Apart from his studies at the academy, he attended what public lectures he could in Lancaster on topics that interested him. A series on astronomy, physical geography, geology, mineralogy, paleontology, and related subjects apparently made a particularly deep impression. He also took a course of lessons in freehand drawing and made good use of the Lancaster Public Library, said to have been one of the best town libraries in the state at that time. It was here that he discovered Darwin; here, too, that he read deeply in the works of Henry David Thoreau (1817–62), a naturalist much after his own heart, and of the German naturalist and explorer Alexander von Humboldt (1769–1859). Humboldt's *Kosmos*, a description of the physical universe in five volumes, has been called one of the great scientific works of the day, and he had had a profound influence on both Darwin and Alfred Russel Wallace. Of him Burbank remarked that "perhaps the basic thought I absorbed was the idealistic and intrinsic worth of the work in which I was later to embark." These three writers—Darwin, Humboldt, and Thoreau—he identified as the decisive "book influences" that "stand out in my life as having influenced greatly my career."[13]

Reading was for the winter. At his summer job, he showed a distinctly mechanical turn of mind. He had a natural bent in that direction. When he was sixteen, he constructed a wall clock that was attached to a beam in his father's house and is said to have kept time with a fair degree of accuracy. He also made a set of dumbbells for the academy gymnasium—he was very fond of athletics—and worked at a series of machines and gadgets, waterwheels, and steam engines. His wages at the plow works were only fifty cents a day at the outset, not enough to pay his board at his uncle's house. Presently, however, he invented an attachment for the engine lathe, which turned out to be so effective as a laborsaving device that, working by the piece instead of by the day as had first been the arrangement, he was soon taking home a fatter pay envelope on Saturday nights. But Luther hated the noise of the shop, and he was allergic to the sawdust that filled the air. He began to look around for other work. Encouraged by his father to think of becoming a physician, he started to direct his studies along these lines.

Around this time, Luther experienced a "religious awakening." He was baptized in the Nashua River—named for those long-gone Nashaway Indians—and joined the "hard-shell" Baptist church at Still River. This seems to have been under the influence of his mother. The Burbank women were always veering off in that direction, taking their children with them if they could. Grandmother Ruth had done exactly the same thing in her day. In 1870, however, Luther and his mother transferred to the less severe Baptist church at Groton Junction. In his later thinking, he gradually reverted to the milder spirit of the Unitarianism his grandfather Nathaniel had brought with him from Harvard and ultimately to the Emersonian transcendentalism that sprang from the blending of Unitarian rationalism with the deism of the eighteenth century.

Luther showed the distinct stamp of his Puritan background. Still in his teens, he came across some lectures on diet and the "water-cure" and adopted the harsh regimen advocated. "Taking a pail of water to his room at night, which would be frozen over so hard in the morning that he would be obliged to break the ice with a stick of firewood, he would take a sponge bath in this ice water," says his sister. "The window of his room was always open and sometimes the snow would drift in upon his bed. About this time he resolved to eat no meat and only a very limited amount of food, which he carefully weighed so as not to exceed the quantity he allowed himself. All the regulations he deemed good he carried out to the letter, until he so reduced himself in flesh and strength that our parents called in the family physician, who objected to the strenuous régime. Luther, however, always claimed that his health was permanently benefited by the experience."

In the summer of 1868, he suffered a severe sunstroke while running to get help to put out a fire sparked by a passing locomotive in a timber lot on the Burbank property. "Luther, then a slight youth of nineteen, was five feet and eight inches in height, but became reduced from his usual weight of 130 pounds to 120 pounds," Emma says. "Failing to recover from the effects of the sunstroke, in August 1870, he made a cruise in a mackerel schooner bound for the Gulf of Newfoundland. The vessel was wrecked, but he was so seasick at the time that being taken on board a lumber raft and towed to shore seemed of small matter to him. He returned home minus his baggage, but greatly improved in health."

It seemed only too clear that there must have been another element to this protracted convalescence of over two years. Our information is scanty, and the notion of psychosomatic illness was still in the future, but an ocean voyage would be an improbable prescription for sunstroke and a likely one for a nervous breakdown. Luther did not return to school and seems also to have given up his job at the Ames Plow Works. He was going through a complex psychological reorientation. The intense, ascetic youth was flowering under the influence of new ideas. The hard-shell Baptist was in conflict with the newborn Darwinian. By no amount of contortion could these world views comfortably cohabit.

An additional strain was the death of his father in December 1868. The life of the family was painfully disrupted. Luther was thrown back on his own resources. He decided to abandon the medical studies he had begun under a local doctor, but he could not dally much longer in the choice of a career.

Darwin's investigation into the processes of artificial selection was fresh in his mind. The emphasis of that inquiry was more on animals than on plants, but the idea of devoting his life to horticulture had begun to form. It was more of a realization of the dramatic potential of plant breeding than an intention, but it was there. In a letter dated January 1909, Burbank subsequently wrote:

> When I was about nineteen, in 1868, probably the turning point of my career in fixing my life work in the production of new species and varieties of plant life was . . . the reading of Darwin's "Variation of Animals and Plants Under Domestication," which I obtained from the library at Lancaster, Mass., my old home. Well do I remember reading that work of Darwin's—that the whole world seemed placed on a new foundation. It was without question the most inspiring book I had ever read, and I had read very widely from one of the best libraries in the state on similar scientific subjects. I think it is impossible for most people to realize the thrills of joy I had in reading this most wonderful work. The reading of this book was without doubt the turning point in my life work.
>
> At once, as soon as I was able, I purchased other books written by Charles Darwin, and today I have still one to read which probably influenced the general public more than all the rest of his writings—"Origin of Species."
>
> I have been so busy producing living forms from the thought inspired by Master Darwin's conclusions that I have never, to the present date, had time to read his "Origin of Species." However, I imagine I could write the "Origin of Species" myself from what I have read of his other works.
>
> Darwin was the greatest man who lived in the past century, and in his thoughtful, truthful, laborious, unobtrusive way, in my opinion, did more to liberate and ennoble the human race than any other man who lived during the century.
>
> The second book which I purchased of Darwin's was "Cross and Self-Fertilization in the Vegetable Kingdom."

In 1869 the family left the old home at Lancaster, selling their half-share in the farm and house in Luther's cousin Calvin, who had inherited the other half on the death of his father, Aaron Burbank. Olive and her children moved to Groton Junction (subsequently known as Ayer Junction), where she bought a cottage on the outskirts of the village. Two years after this, Luther learnt that "seventeen acres of unusually fine soil" were for sale in Lunenburg Township, not far from Lancaster. With the small inheritance that was his share of his father's estate and a mortgage to make up the difference, he bought this piece of land, which had on it an old house in good condition. His mother acquired a house in Lunenburg too, with land adjoining Luther's, and moved there with Emma and Alfred.

On these seventeen acres, Burbank began truck gardening, producing vegetables for sale at the market in the nearby town of Fitchburg. He also sold sewing machines door-to-door "up Fitchburg way," canvassing with a horse and wagon. "I was not naturally a good salesman," he later wrote. "I never quite got over the dislike I had of driving into a farmer's yard and leading up to the fact that I had come to sell them something. After I got to showing off the machine, I rather enjoyed the work, but each call was a separate test of my 'brass,' which I felt I lacked. . . . I was careful to keep in localities where I was not known, as I felt I could sell better to strangers, and I knew it would never be my life work, so I did not then wish to become identified with the enterprise."[14] Though the pay was good, he presently gave up this work and turned back to the task that was to make him famous. It was, he said, more cold necessity than any particular intention that showed him the way.

 5

BURBANK'S SEEDLING

Darwin had lit the fuse; circumstances were the charge. It was the very challenge of those early years that laid out the lines of Burbank's career. He was now obliged to compete with established vegetable producers, which was hard going unless he could somehow gain an edge on them. He learned the secrets of his new trade—the uses of cold frames, hotbeds, and fertilizer—and he learned them thoroughly, gaining a professionalism that would stand him in good stead all his life. He was in competition with more experienced men, long set up in business. By cutting corners in production and marketing, he could produce vegetables more cheaply than other gardeners, perhaps, but what really fascinated him was the possibility of raising better vegetables and getting them to market ahead of the others. "Here," he thought, "was a place to use wits and ingenuity and inventive faculty and another faculty that, at the time, I did not know I had, even latently—the power of choice as between two apparently identical varieties or even between two apparently identical plants!"[1]

Thinking about the ways in which nature produced variations in plants, he remembered the "winter garden" he had found while wandering in the Lancaster woods as a boy, where he had seen summer plants growing amid the snow, nourished by a deep-flowing spring of warm water.

> I knew enough to know that vines and shrubs and flowers grew according to their season and withered according to an hereditary tendency that had impressed itself on them through thousands of generations of growth. The warm spring accounted for the equability of the temperature in that spot, but it did not account for the luxuriant and lush growth of plants that should be dead. What I had supposed was a fixed and immutable law of Nature was not, then, fixed and immutable at all. Or, better, the law was there, but I had misinterpreted it. What was the influence behind this phenomenon?[2]

He began to experiment—in what he admits was "a crude, superficial way, without any order to my trials and without keeping any record of results." He noted something every gardener has noticed: that certain plants grew faster than others in the same row. Later in the season, he marked out those that bore more or bigger produce than ordinary or fruit that was of superior quality or ripened earlier or later than average. Some plants were actually superior in more than one respect.

Individual plants "willing to take a step forward" (as he put it) were marked for saving come harvest time, and their seeds were kept for the next year's planting. At first he was far from sure that the improvements would be reproduced in the next generation, "but after a time I began to see that they were, and that there was certain to be at least part of the second year's growth that would show all the finer characteristics of the marked plants, and a tendency to improve on their improvement."[3] Slowly he began to believe that he was on the right track.

Creative horticulture was still very much a sideline, taking second place to his market garden. (He kept a few animals, too—a heifer calf of his winning the second prize, a dollar, at the Lunenburg Town Fair in September 1874—and sold honey to bring in extra income.) But he persevered at it. In the beginning, he used seedsmen's collections, testing, selecting, and crossing those that seemed most promising. He also experimented with various wild flowers, gaining an acquaintance with them that would later prove useful (particularly in the development of the Shasta daisy, in which Japanese, European, and wild American species combined to produce a unique hybrid that holds its own even today). He did some work with various kinds of beans and was able to produce a variety of sweet corn that came in a week earlier than ordinary. "It may be said that in general my garden products were of exceptional quality," he later asserted of this period.

One early failure was an attempt to produce a yellow sweet corn. In those days, all sweet corn was white, and yellow corn was thought to be fit only for animals. Burbank cross-pollinated white sweet corn by yellow field corn in an attempt to fix a yellow sugary kernel. But he was unable to get a uniformly yellow ear unmixed with white seeds while he was at Lunenburg. There were some promising hybrids, but when he moved to California the project was abandoned, and he did not resume his experiments with corn until much later. At this stage, he does not appear to have realized that it was necessary to continue breeding beyond the first generation. Like other breeders, he was working very much in the dark. Ironically, as the geneticist Donald Jones (himself one of the outstanding figures in the corn-breeding program that revolutionized American agriculture) pointed out: "Thirty-six miles away, in the library of Harvard University, were the Proceedings of a Natural History Society in Austria. In that publication was a clearly written report, printed a few years before, by a Catholic priest that would have told Burbank exactly how to fix his yellow sweet corn in the shortest possible time. But it was written in German and the pages were uncut."[4] Those pages would remain uncut for another quarter century. Hugo de Vries, who would be among the first to understand their significance, was still a postgraduate student in Germany. Mendel's time had not yet come.

Burbank's first notable achievement, and the point at which his lifework really began, was the development of the Burbank potato. This called for no understanding of Mendelian law, inasmuch as it was nothing more or less than an amazing stroke of good fortune: the improbable sort of opportunity that proverbially knocks once and disappears forever unless unhesitatingly grasped. Burbank was ready to grasp it. Nothing else he ever did would be as successful in the long run as this potato.

Potato cultivation in South America may go back to before 3000 B.C. In the Andean highlands, it was the most important single food plant in pre-Columbian times, and more than ninety wild species are still to be found growing there, along with several hundred cultivated varieties. During the latter part of the sixteenth century, Spanish explorers brought potatoes back to Europe, probably from Chile or Peru (the exact date and provenance of its introduction into the Old World are unknown), and via Spain they were carried back across the Atlantic to Florida, whence in turn they reached Virginia. In 1586 Sir Thomas Herriot, a companion of Sir Walter Raleigh's, apparently brought the first potatoes to Britain from there. They did not find much favor among the English of that day, though Raleigh tried to interest Queen Elizabeth and managed on one occasion to get a dish of potatoes served at the royal table. Not until the middle of the seventeenth century were potatoes widely grown in England. On the continent of Europe they met with even more resistance. To encourage their use, Louis XVI of France ordered potatoes planted and went so far as to wear a potato flower in his buttonhole. The fields that had been planted to potatoes were guarded by troops with a view of provoking public curiosity, and the king served banquets in which the tubers featured prominently. None of this seems to have overcome the suspicious resistance to the new vegetable. Only in Ireland was the potato favorably received, by an impoverished peasantry that could not afford to be persnickety about its diet, and the first cultivation of potatoes in northern Europe took place in that country in the vicinity of Cork. Perhaps it was because of its relationship to the deadly nightshade (which belongs to the same family, the Solanaceae, as do the tomato, tobacco, and the eggplant) that the legend that potatoes were unwholesome arose and persisted so long.

When finally accepted, the potato soon demonstrated its remarkable properties as a food. Seven pounds of potatoes and a pint of milk—the basic diet of the Irish peasantry in the eighteenth and nineteenth centuries—are said to supply all of a man's basic nutritional requirements.[5] Potatoes yielded a significantly larger quantity of food per acre than other staples. They must have seemed a miraculous answer to the problem of famine that had plagued Europe since the beginning of its history.

In the 1840s, however, potato blight, caused by the fungus *Phytophthora infestans*, first appearing in Germany, destroyed the Irish crop year after year, and in the next two decades an estimated million people died of starvation in Ireland, where widespread dependence on this single plant combined with official indifference to produce a long-drawn-out famine. The blight also ravaged the fields of America during these years, and in 1859 the situation was further aggravated by the appearance of great numbers of potato beetles in Colorado. Native to the area east of the Rockies, where they live on wild plants of the potato family, the beetles spread rapidly through Nebraska, Iowa, Illinois, and Ohio, and soon reached the northeastern states. At the same time, a mysterious tendency on the part of the tubers to degenerate manifested itself. It seemed an end to potato growing.

The cause of the potato blight was unknown, and it was regarded as part of a degeneration syndrome now attributed to an accumulation of tuber-borne viruses within the vegetative clones.[6] It was thought that the potato had degenerated as a result of being propagated exclusively from cuttings, and though other factors were recognized—lack of care in selecting seed potatoes true to type and storage under conditions that reduced vitality—horticulturists believed that the plant needed sexual reproduction (that is, propagation by seed) to restore its vigor. In 1851 C. E. Goodrich of Utica, New York, procured potato varieties from Panama (the U.S. consul simply bought them for him in the marketplace there), one of which he misleadingly called the Rough Purple Chile. In 1857 another type, the Garnet Chile, was developed from this, which in turn gave rise, in 1861, to the Early Rose potato.[7] This, though an improvement, was generally speaking small, reddish-colored, and kept poorly. The agricultural papers made much of the need for better potatoes. "It required no genius to know," Burbank later observed, "that if a large, white, fine-grained potato could be produced, it would displace the other varieties and give its discoverer a great advantage over his competitors." With this in mind, he attempted crossings, but the blossoms produced no seed. And since he was dealing with a cloned species, reproduced from cuttings that must give rise to progeny identical to the parent, efforts at selection were pointless.

Then he had a stroke of unbelievable luck. He discovered a pod of seeds ripening on an Early Rose potato plant. Such seedballs are not unknown, but they are very rare. In later years, Burbank was to offer a standing reward for one, but as long as he lived he never saw another. He watched the growth of this solitary seedball day by day. One morning it had disappeared, and he was devastated. Then it struck him that perhaps it had been dislodged by some dog running through the patch, and he made a careful search of the ground. The seedball was soon found a few feet away from the parent plant. He carefully preserved and planted the seeds. There were twenty-three of them, and he raised twenty-three seedlings. Though a number of these, he says, were better than any potato then grown, he selected only two. They were as different from the Early Rose, he asserts, as modern beef cattle from the old Texas longhorns. In this alone, he was lucky. Like other species propagated by cloning, the potato has a great tendency to revert to its wild form when grown from seed. It is often necessary to cultivate hundreds or even thousands of seedlings before any useful new type is produced.

He was twenty-four when the first Burbank potatoes saw the light of day in his Lunenburg truck garden. "It was from the potatoes of those two plants, carefully raised, carefully dug, jealously guarded, and painstakingly planted the next year, that I built the Burbank potato," he says in his autobiography. (Walter Howard avers, however, that "all but two of the clusters were worthless. One of these was later discarded, while the other was kept and multiplied as rapidly as possible.")[8] It seems probable that the reference to "building" the Burbank potato was the work of Burbank's ghost-writer, Wilbur Hall. Further hybridization may conceivably have been involved, but in all likelihood the new potato was simply the chosen seedling and its progeny.

He could not in his wildest dreams have imagined how successful it would be. A presumed "sport" of this original seedling, known as the Russet Burbank, which is nearly identical to it but for its russeted skin, is the most widely grown potato in the United States at the time of writing—over a hundred years later. Descendants of the Russet Burbank (also known by the names Idaho Russet, Golden Russet, and Netted Gem) are also prominent, among them the Early Gem, Shoshoni, Reliance, Norgold Russet, Targhee, Nampa, and Nooksack. (Because of its complete male sterility and high female sterility, the Russet Burbank has been little used in breeding new varieties, but new techniques have been developed to overcome some of the female sterility and more use of it in breeding is now being made.)[9]

Burbank's own profit was largely indirect. In 1875 he sold the whole stock of the potato to James J. H. Gregory, a seedsman of

Marblehead, Massachusetts, whom he happened to know. ("My Father taught school in Mr. Burbank's home town and was acquainted with him," his son Edgar Gregory afterward noted.) Gregory paid Burbank $150 and allowed him to retain ten of the tubers for his own use. He also did him the considerable favor of naming the new potato Burbank's Seedling.

Later on Burbank would attack the potato again, crossing common cultivated varieties with *Solanum maglia,* the Darwin potato, a wild species from Chile; with *S. commersonii* from Argentina; with *S. jamesii,* the squaw potato of the Southwest; and with an unidentified Mexican species. In 1895 he crossed the Pacific Coast variety Bodega Red with the Burbank to produce a hybrid of which eight varieties were advertised, but never introduced. Writing in 1912, the Colorado grower E. H. Grubb indicated that he and Burbank were planning an expedition to the island of Chiloe, off the Chilean coast, believed to be one of the original sources of the potato. This trip was never made, however, at least not by Burbank, who makes no mention of plans for it in his writings.

The high degree of luck that went into the finding of Burbank's seedling is evident from the fact that he was never afterward able to do anything remotely as important with potatoes, despite all his efforts. At the same time, the fact that the new variety bore his name did much to lay the basis for his reputation. Fate, in leading him to that original Early Rose seedball, had undoubtedly smiled on him.

"Perhaps one of the first words I ever heard was California," Burbank remarks in his autobiography, in a somewhat rambling dissertation on the whys and wherefores of his move west. Early in 1847, gold had been found in the race at Sutter's Mill on the American River, sparking a frenzy of interest and activity. The newspapers proclaimed the discovery of El Dorado, and from then on emigration to California proceeded in a flood from all parts of the country, notwithstanding the great hardships involved in getting there. According to official census figures, the population of the state increased from 224,000 in 1852 to 380,000 in 1860: an increment on the order of almost 20,000 immigrants a year. Two of these were Luther Burbank's much older half-brothers. George had gone to California in 1854, and David followed him in 1859.

They had had to make the difficult overland journey by horse and wagon, five or six months' travel beset by illness, Indian troubles, and endless exertion.

On May 10, 1869, however, the final, golden spike of the transcontinental railroad was driven home, and this same journey could henceforth be made in nine days, still in relative discomfort, but at any rate safely. In the summer of 1874, Luther's twenty-two-year-old brother Alfred followed George and David. He found work in the town of Santa Rosa as a carpenter. The temptation was more than Luther could longer resist. "I had no choice in the matter," he says. "I was bound to go!"

It is hard to imagine any ambitious young man not feeling the lure of the West then. "The forty years that lay between the California Gold Rush of '49 and the Oklahoma Land Rush of '89 saw the greatest wave of pioneer expansion—the swiftest and most reckless—in all our pioneer experience," Vernon Parrington remarks of this period in American history.[10] And Burbank—who was of an age with that period and had grown up amid the great hullabaloo of expansion and exploration—was bound to be a little swift and reckless too. "Within sixty days of the time when the definite decision to go to California was reached," he tells us, "I had sold my personal property and closed out my business at Lunenburg."[11]

But there was another reason for his going so far and so suddenly: he was apparently suffering from a bad case of unrequited love. Her name was May, or Mary, Cushing. There had been a disagreement. She was less ardent than he and made it clear to him that he was not the only eligible young man around. (After his arrival in California, Burbank wrote to her, and their friendship was reestablished by mail, evidently lasting to the end of his life.) A broken heart; the limitations he felt in Massachusetts, where the clergy still railed against Darwin; the lure of the Golden West; and a sense of the potential of California as a horticulturist's paradise thus all combined to contribute to his departure from his native New England.

Burbank was twenty-six—in his own words, "a small, wiry, active young man, observant, alert, inquisitive, and full of ideas about what I was going to do." As yet he had seen little of the world outside rural Massachusetts, where his amusements had been the village singing school and the choir of the little Methodist Episcopal church that stood near the Burbank home in Lunenburg. He had no small opinion of himself, however, and after his success with the Burbank potato, his future vocation seemed clear. He went with a purpose: to establish himself in the profession of plant breeding in the new lands by the Pacific. With his land, farming tools, and stock disposed of and his mortgage paid off, he was worth just $660 (which included the $150 James Gregory had paid him for the potato).

The age he set out to woo was tough-minded and ruthless on the one hand, sickly sentimental on the other. Unscrupulous exploiters of the stamp of Cornelius Vanderbilt, Jim Fisk, and Jay Gould bestrode the land—as did their more lowly counterparts the James boys and the Younger brothers. But, thought Henry James, it was for all that "a feminine, a nervous, hysterical, chattering, canting age, an age of hollow phrases and false delicacy and exaggerated solicitudes and coddled sensibilities, which, if we don't soon look out, will usher in the reign of mediocrity, of the feeblest and flattest and most pretentious that has ever been."[12]

U. S. Grant was president, and his eight years in the White House, Vernon Parrington says, "marked the lowest depths—in domestic affairs at least—to which any American administration has fallen." The nation was coming up for its one hundredth birthday. In Philadelphia, second largest city in the United States, preparations were under way for a great centennial exhibition. It was a time of expansion, of public corruption and private gentility, of labor strife and mass unemployment. "Everywhere," says Parrington, "was a welling-up of primitive pagan desires after long repressions—to grow rich, to grasp power, to be strong and masterful and lay the world at its feet. It was a violent reaction from the narrow poverty of frontier life and the narrow inhibitions of backwoods religion. It had had enough of skimpy, meagre ways, of scrubbing along hoping for something to turn up. It would go out and turn it up."[13]

Into this world went Luther Burbank—slight, something of a hypochondriac, perhaps a bit of a mother's boy (even Alfred had struck out before him). He, too, would go out and turn his something up.

6

"THE CHOSEN SPOT OF ALL THIS EARTH"

On the strength of the sensational Ralston bank failure of August 1875* and various reports disparaging conditions in California, friends tried to persuade Luther not to go. But his mind was made up, and all the preparations were complete. He would not back down now. In the latter part of October 1875, probably on the twentieth, he set out.

With natural Yankee parsimony, he had decided that he could not afford a sleeping berth, which was in any case a somewhat unusual luxury in those days. He was therefore obliged to sleep curled up on the seat throughout the nine days and nights of the journey. When, on rare occasions, he found himself with a whole seat to sleep on and could stretch out, the trainmen moving up and down the cars always seemed to be banging into his projecting feet. This remained one of his most vivid memories of the trip.

His mother had provided a basket of food, which proved very necessary, as the train was sometimes held up a whole day far out on the plains, out of sight of even the smallest settlement or homestead. The axle boxes got so hot that long stops had to be made for repairs. Burbank several times shared his lunch with passengers who were less well provided for than he.

* The leading financial institution on the West Coast, the Bank of California, of which William Chapman Ralston was a founder, failed in August 1875 with substantial losses to depositors. See George D. Lyman, *Ralston's Ring* (New York: Charles Scribner's Sons, 1937).

He wrote to his mother and sister Emma back in Lunenburg virtually every day—even if it were only a postcard. Emma, who was quite ill at the time, saved some of this correspondence, which gives a good idea of the journey and the young adventurer:

> [Between London, in Canada, and Detroit, Mich.]
> Have been going westward steadily since I left Worcester at a great speed. No stops longer than twenty minutes. Slept some last night. Have enjoyed every mile. Had a treat in crossing Suspension Bridge by moonlight. Saw Niagara and the rapids. Cars joggle me. Want to hear how Emma gets along as soon as possible. I wish you were all enjoying the ride and scenery with me. We are in Queen Vic's dominions. Mrs. Ward and son have stood it nicely without any trouble. Don't know where I can mail this. The sandwiches and cake are tip-top. The cars are vastly more comfortable than those on the short Eastern roads.
>
> The forests are too beautiful for one to describe—zigzag fences. Saw Lake Ontario at Lockport last night. Will write again before long.
>
> *Lute*

> [Card from Burlington, Iowa]
> Just crossed the Mississippi. Have had extra good luck in making changes, enjoying myself. Seas of corn, thousands of wild ducks.
>
> *L.B.*

> [Pacific House, Council Bluffs, Iowa]
> Dear Mother, Etc.:
>
> I arrived here at 10 o'clock last night. The passengers all stopped either here or at Omaha last night. We start for Omaha from here at 9:30 this morning. I have enjoyed myself, been perfectly contented and have taken more comfort than a little in the journey. It has been perfectly delightful except changing cars at 11 o'clock at Chicago when I was sleepy. Have met a number of ladies and gentlemen who are on their way to San Francisco. Several of them stopped at this house last night. There are three ladies and two gentlemen from Fitchburg who have come all the way with us. All the people on the cars are pleasant, sociable, and obliging. All *have to* enjoy the scenery. The babies make some noise at times, but I can settle down in a seat and go to sleep amid all the noise and sleep as sweetly as I ever did in my life. We crossed the Mississippi into Iowa yesterday morning at 8 o'clock. It took us all day till 10 o'clock last evening to cross the state, and we went fast, too. Have seen a few Indians, but no Chinamen yet. Have eaten all my sandwiches, one-half of cake. Shall have enough

to carry me twice as far as I have to go. I buy coffee twice a day. It is very nice. Should like to hear from home now. Hope Emma is getting better fast.

This letter is a curious one, but you must take it for what it is worth.

Yours in haste,
L.

[Card from Cheyenne, Wyoming Territory]

Near the Black Hills.

Have been travelling since 8 o'clock this morning without seeing a tree or bush over one foot high. A dusty desert. The passengers are enjoying themselves looking at prairie dogs, antelopes, and at the stations hunting for moss agates. The prairies were on fire last night. A finer sight I never saw, but called by the settlers "Prairie Demon." Wish you could only see it.

Just went through the first snow shed.
Near Pike's Peak.
6040 feet above you—more than a mile!

They are the ordinary communications to his family of an apparently ordinary young man, making a trip that was already (it was six years since the opening of the transcontinental line) fairly commonplace. No one would suppose from them that young Lute, with his "tip-top" sandwiches and cake from Mama's hamper, and coffee twice daily, carried with him any particular spark to ignite the new land.

He reached San Francisco late on a Friday afternoon. By Sunday he was in Santa Rosa where brother Alfred was settled. That same day, October 31, he dashed off the promised letter home:

Here I am in Santa Rosa, and before I give you a view of this place I will give you a glance at the journey across. . . . We changed cars at Worcester, Utica, Rochester, Suspension Bridge, Detroit, Chicago, Council Bluffs, Omaha, Ogden, Oakland, San Francisco, and Donahue, but after Omaha the changes are so far apart that they are rather pleasant than otherwise, though it would be very hard for a woman if she had any baggage to look after except what she might carry in her hand.

. . . Got acquainted with Judge Miller of Chicago (his son expects to settle in this town), also a wealthy Michigan banker's wife and *daughter!!!* The journey to Nebraska Center was pleasant. After that it is all *desert* for more than a *thousand* miles, and with one exception I saw not a tree the whole 1000 miles. Some of the way, in fact a good share of it, is *more sterile than a bare granite rock*, with poison water, mines, coal beds, salt, and alum springs and every other nasty chemical. Someone made the remark that they thought it must be the *roof* of H——. If you saw it you would see the point. Through the desert there is nothing of special interest except inside the cars, and there the dust *gathers* in drifts, especially in a fellow's nose and throat, which it irritates terribly.

. . . The rest of the way is *delightful beyond description*. Mountains piled on mountains, snow-clad peaks gilded with sunlight, wild forests, rivers, deep cuts, mammoth trees, leagues of snowsheds, thousands of geese, ducks, swans, antelopes, etc. *I cannot give you any idea of them*. Then on the Sierras, when we were winding up the mountainside, could look down thousands of feet onto the *lovely* Donner, Honey, and Tulare lakes. Then we rushed without steam, except to hold the train back, down the Sierras into the golden land where the old gold washings are. It looks as if there had been two or three earthquakes and a flood or two.

The train stopped a few minutes at "Cape Horn," where we got out and looked *down one-half mile* into the valley. After riding a few miles further gardens appeared with olives, oleander, fuchsias, figs; and all kinds of fruit trees and vines are green and thriving. On down into the great Sacramento Valley, where the gardens and front yards are ornamented with palm trees, century plants, fig trees, etc.—*it was a rare feast for me*. Then on across the valley as level as a floor and as *rich* as mud to Oakland, which is a beautiful place. At San Francisco I stayed two nights and one and a half days. I cannot describe the joys I felt in looking at the gardens and feeling the healing balmy breezes, but liquor-selling is the great business of that great city. No one who has not seen it can imagine the amount consumed. I used to go nearly one-half mile to get a glass of good water, and I knew of only one or two places where it could *be got. I made a vow on my way over that I would not touch a drop of any kind of liquor, and I shall keep it*. Of those who do not drink there are a great many, and they are almost *without exception* the *leading* and most *respected* men, and who also own most of the property and do the important business. A young man who will not drink here and is good-natured and makes folks like him, and who *minds his own business*, has *ten thousand* chances of success where the *same* qualities would have *one* chance in the states.

Now you want to know how Alfred appears to me. His chin whiskers are grown out so he looks some different in that respect. Then he looks more rugged, but is not quite so fleshy as he was a while ago, but is in pretty good condition. The thing that struck me most about him is his great haste to get rich, which thing never bothered him when East. He is in for making money. He is out of a job now and is a little blue—is afraid he cannot get one through the rainy season, which, by the way, is expected every day. He and another fellow have put up an 8 by 10 shanty. I expect to go to keeping house with them tomorrow. We bought crockery, bed ticking, etc., last night. Dave and Lina are coming up after me soon. I shall not look for work for a week. The change of climate has given me a cold, as it does nearly all, but I never felt so *contented* and free from mental disquiet and never slept or ate better in my life. (There are two fellows in the room talking, so please excuse blunders.) There are some Chinamen in this place. I like them very well. They know about four times as much as folks generally give them credit for. They are disagreeable in some respects.

I want to know how Emma is getting along very much. Expected a letter before this. I agreed to write all about things when I got here, so I began today, and I have given you as good an idea of things as they appeared to me as I can, except a description of this city, valley, and surroundings, which I will give you on another piece of paper.

Love to all inquiring friends.

Luther

His luggage when he arrived contained little aside from his clothes, books, and a supply of garden seeds, including the ten Burbank potatoes Gregory had permitted him to keep. Having chosen Santa Rosa (he had two elder half-brothers living at Tomales, but it seemed to him that that was too close to the ocean and that the climate would be unsuitable to his experiments), he was determinedly enraptured by the place, then a small village surrounded by great fields of wheat. Designated county seat in Sonoma County by a vote taken in 1854, it dated from a settlement founded in 1833 by General Mariano Vallejo. It was, Burbank declared, on the separate sheet of paper, included with the preceding letter, "the *chosen* spot of *all this earth* as far as *Nature* is concerned, and the people are far better than the average Californians in other places." This last sentiment must have been taught him by Alfred and his friends: he himself had, after all, had scant opportunity as yet to become acquainted with the other places in question.

The cost of living, he was sure, was half what it was in Massachusetts. "Meat costs but little, flour is better and cheaper, fruit is nothing, almost, very little fire is needed, and such warm expensive houses are not necessary." He tacked on a postscript, with further raptures:

> A fog is hardly ever seen here—the wind never blows hard. I wish you could see California fruit. I bought a pear at San Francisco, when I thought I was *hungry*, for five cents. It was so large that I could only eat two-thirds of it. I threw the rest away. Grapes are so abundant that all are allowed to help themselves to the nicest kinds at the vineyards. There is no skin to them and very small seeds; the pulp is the whole grape. If you try to squeeze one out it will split like a plum. They are very sweet and nice and are so plentiful that they are often used as hog feed.

Quoting this letter half a century later, he was obviously a trifle embarrassed by this excess of enthusiasm: "The italics were those of my youth: also the hand-made abbreviation for San Francisco." (In his own version the city is called "S. Frisco," but Emma appears to have tidied up the letter in her rendering.) The "8 by 10 shanty" put up by Alfred and friend proved a durable edifice, in fact and in legend. It survived into the twenties, serving its later owners as shed and chicken coop. The first book eulogizing Burbank, Harwood's *New Creations in Plant Life*, published in 1905, drew from this the assertion that "he had no place to sleep nights, and for months made his bed in a chicken coop, unable to get enough money ahead to pay for regular lodgings." Of such Horatio Alger touches, irresistible copy to the newsmen of the day, was the Burbank mythology, which did so much to obscure his real merits, painted up. Harwood went on to describe the young Burbank "reduced to absolute want" and going to the village meat market to "secure the refuse bones saved for dogs, and get from them what meat he could." In fact, as Burbank afterward admitted (confessing a reluctance to spoil "so beautiful and inspiring an error"), he had not begun life in California sharing his roof with the chickens and had been well-fed and happy at the time.

The separate section of the letter describing his new home was marked "Not Public"—and the explanation (which Emma tactfully omits) was as follows:

> The reason that I give this description of Santa Rosa outside my general letter is because if it is generally known what a place this is all the scuffs would come out here, get drunk and curse the whole country, so don't let on to *anybody* outside the house what sort of a place this is *except* that I am delighted with it.[1]

But if he did not live in a chicken coop and eat dog meat, all was nevertheless rather less rosy than he anticipated. The rains had begun, and there were few jobs available. The immigrant's life was a hard one, for all the idyllic surroundings. In the summer, there was work in the wheat fields with the harvesting and threshing crews or breaking the soil with gangplows pulled by teams of oxen or mules. But Luther's physical strength was not up to such strenuous labor. In the winter, there was some construction in town, but that was not the work he wanted either, and there were many applicants for every job. The coming year was to be one of the hardest of his life.

"So [he says] I found myself almost without means, in a strange land, far from home and friends, and there was no obvious way in which to enter on the specific work that was contemplated." The immediate problem was simply a matter of earning enough to support himself, and he could not afford to be choosy about what he did. Life in the shanty establishment was cheap enough: $1.94 a week each, they reckoned, was what their simple housekeeping cost them. They got their supplies in bulk: "a sack of sweet spuds, one sack of onions, one sack apples, ten pounds oatmeal, some fresh canned salmon, etc. etc.," he informed Emma in a letter home.

Since he had paid $140 for his ticket to California and had spent only an additional six dollars for "coffee, tea, bread, lodging at hotel, and *all*" en route, Burbank should logically have had quite a bit left of the $660 he had started with. But this nest egg he was presumably saving against the day when he would be able to enter on the career he planned for himself. In the meantime, along with the other young immigrants, he eagerly went after whatever odd jobs offered. He was never one to live on his capital. Once, hearing that there was work available on a building that was being put up, he learned on applying that he could have a job if he could supply his own shingling hatchet. He went out and bought one, spending his last spare dollar in the process, he says, but was bitterly disappointed when he went back to find that another man had been taken on in the meantime.

He found time to do some exploring and hiked the countryside all around Santa Rosa. In one place he discovered "enough new and curious plants . . . to set a botanist mad. There is an old surveyor who knows nearly all of the plants here. I am going to take a batch to him this evening. . . . My botany tells the names of only a few California plants. Some of them *have no names*."

All the while he continued to sing the praises of California's agricultural wonders in his letters home, his postscripts, and post-postscripts—they are replete with descriptions almost too extravagant to be believed:

> There is a large field of squashes here in which there are probably five hundred that two men could hardly lift—some of them three and a half feet long by one foot three inches through. Hundreds of them bigger than the biggest ever seen in Lunenburg. I asked Alfred why he didn't write home about the climate, squashes, and big things. He said he *dare not*. *I will* run the risk of being called a whopper teller,

"Bro. Lute" informed his family, and suiting the word to the deed immediately went on to extol in a postscript:

> Bunches of grapes half a yard long!!! The white Muscat or Alexandria grape—three and one-half by three and a quarter inches in circumference; if you have a Lunenburg grape to compare, you will observe the difference.

There was a typical young man's desire to impress the folks back home. With this letter (November 9, 1875), he included a Chinese laundry receipt. "Can you read it?" he asks. "I should like to see the white man that could." But he had a natural tendency to exaggerate. He was of a kind with the scouts who sent back the first reports of the land of Canaan and, like those Israelites, had the two-handed faculty of making his public believe him. It was not that he was lying. He simply *saw* a little larger (elsewhere he reports "a field of cabbages nearly every head of which was as large as a washtub") than life, particularly when it came to growing plants, which loomed up so vividly in his mind in the first place. But in later years tall stories would get him into any amount of trouble with hard-headed scientists busy ordering the universe under the banner of exact measurement.

He found some work lathing at the new Palace Hotel, the largest building in Santa Rosa, but this was only good for about a week, as quite a number of other men had been taken on for the job as well. Santa Rosa was growing apace. Where wheat had been harvested the previous summer, he told his mother, houses now stood by the hundreds. The recent arrival of the railroad had opened up the valley to development. The lathing job he had expected being delayed, he decided to pay a visit to his brothers at Tomales. He spent several weeks with them and, he reported,

> had an outrageous good time, but it would take me three months to write all about it. I went to the ocean, San Rafael, to the neighbors, to a party, to several pleasant rides, sawed wood, milked, looked around the country, and enjoyed myself hugely, and got fat and *good natured*. Everybody in California has a "happy don't care" look and . . . everything is off-hand and original, and I verily believe that I am pretty well Californianized. Folks are, however, sometimes awful blue here, and when they do have them it goes bad. I have seen several newcomers in various stages of the blues, but that disease, the blues, has kept so far from me as the East is from the West or light from darkness. It seems as if I had crawled out of an old dirty dungeon into morning sunshine, but of course that is not all California, but partly circumstances.

He seems to have spent most of his time with David and his wife, Lina. George was in a hurry to put the finishing touches on his "magnificent house" before going up to the state legislature, of which he was a member, on December 6. While at Tomales, Luther investigated Petaluma and San Rafael as possible alternatives to Santa Rosa. At Petaluma he and David searched around for some small farm that he might either hire or buy. "Land about the place on which *anything* in *creation* can be raised can be got at $100 an acre," he says. "Around Santa Rosa for $300 an acre." Nonetheless, he did not find anything that suited him and shortly returned home to bachelor life:

> When I came home last night Alfred said Mr. Walmsley had not cooked anything or done anything during the two weeks that I was gone. Mr. W. told exactly the same story. I believe all that both said and I'll be darned if the shanty didn't look more like a *hog's nest* than anything I had ever seen before, except the real thing.

While Luther was away, Alfred had found a few days' employment at odd jobs. Sick of carpentering, he now proposed to go to work "knocking out paving stones." For his part, Luther decided that he would not take any job lasting more than a few weeks: he was eager to rent his own place and set up for himself. But the rains had put a stop to all work. Many immigrants were returning to the East because there was nothing for them to do. "Mr. Walmsley is going to leave *tomorrow*," Luther wrote his mother on December 5. "Alfred is going to board nearer his work. Mr. Hayes left some time go and a new set are going to occupy the shanty (so our home is busted)."

Luther's hope now was to find a place where he might earn his board. "After fishing two days like a drowning man for a little piece of work about this place, I found myself nearly 'dead broke' or 'strapped,' as they call it here. Had nearly made up my mind that there is *no place* in the great machine that I was fitted for." The immigrant blues seemed to be catching up with him, but he plugged away manfully:

> Thursday morning I decided to make a strike into the country toward Petaluma. After doing up the housework I started (7:30 A.M.). The first ten miles of the way was through *adobe* land. The mud, or rather *mud-slush*, was knee deep to a giraffe. As I had no boat with me, I walked on the fences where I could. Stopped at several farm houses to get a chance to work for my grub till the rains were over, but found no such chance to make my fortune, so I trudged on by the roundabout Sebastopol road toward Petaluma, where I arrived at 6 P.M., a little more than nineteen miles. I thought I would sleep in a haystack, but could find none near Petaluma, and had to put up at a hotel. (By the way, I had nothing to eat after breakfast except a very small slice of bread.) The next morning I got up and addressed myself to breakfast. There was such a panic among eatables about that time as I never saw one man produce before.

After breakfast he looked around the carriage and cooper shops in Petaluma and then applied to W. H. Pepper, owner of the Petaluma Greenhouses and Nurseries. Pepper protested that he had too much help already and that he was going to have to let some of his men go, but Luther persuaded him to take him on,

> and he would give me $50 a month or $30 and live in his family. I thought that was a pretty good offer to *commence on*, at least it was a

little better than running into debt for grub. I think the work is just what I was *got up for*. He has large greenhouses, two great nurseries, a fruit farm, and a ranch. Besides, he has two splendid houses in which he lives, sometimes in one, sometimes in the other.

Saw George Burbank in Petaluma a few hours later. He says I am a lucky bugger. Now *if* I can suit Mr. Pepper and *if* the job don't fall through before tomorrow morning, and *if* I can stand the work, and *if* and *if*, etc., etc., and I have "struck ile" at last, but things are uncertain in this state. Fortunes made and lost in a day.

7

PLUMS FROM YOKOHAMA

He spent the winter and the spring of 1876 working at the W. H. Pepper nursery near Petaluma. With the hundredth anniversary of the signing of the Declaration of Independence there were festivities through the country that year. A general amnesty was granted all unpardoned Confederates, and on May 10 the centennial exposition was opened at Philadelphia by President Grant. One sensation at the exposition was the telephone invented by Alexander Graham Bell. Bell, a mere two years older than Burbank, had come to the United States from Britain in 1872 and was a professor of vocal physiology in Boston, where he had developed his amazing new device. The world was moving forward, and other young men were accomplishing great things. And where was Burbank? Earning his keep as a common laborer and obliged to live in a room over a steaming greenhouse, where he sickened and came down with a fever that nearly ended his life. He went back to Santa Rosa, where he laid up in "the crude bachelor cabin of a workman." Though he could not pay, a neighbor supplied him with fresh milk from her cow. But for this, he said, "it is doubtful whether I should have pulled through. These were indeed dark days."

How was he to get his start? He could have borrowed from his brothers, but he was averse to doing this "both from an inherited sensitiveness about money, which is almost as universal a New England heritage as the Puritan conscience itself, and because I knew that my relatives, in common with such other people as knew of my project, were skeptical as to the practicality of such experiments in plant development as were contemplated."[1] Even to people who accepted the Darwinian theory of evolution, the idea of altering the form of living things seemed grotesque. The degree to which cultivated plants were human creations was not generally recognized in those days. Even botanists familiar with the history of plant breeding emphasized the fact that plants had been under cultivation for millennia and doubted that significant improvements could be accomplished in a single generation. Those who knew of Burbank's plans regarded his project as half-mad and predestined to fail.

Burbank allowed his ambitions free rein. He had plans to improve ornamental trees and shrubs, to look into the possibility of breeding better lumber trees, to produce finer varieties of flowers, and to give the farmers and gardeners of the world a whole range of earlier, sturdier, more productive fruits and vegetables. Everywhere he looked in the fields and gardens around him, he saw potential for improvement. What were his assets? He felt he had a pretty good background in scientific reading. He had been interested in plants all his life and had behind him ten years of practical experience working with them. He thought he had the fundamental rules of plant breeding pretty clearly worked out. Even so, he sometimes wondered: "How could one man, in his single lifetime, have much of an influence on the vegetable world, when about all that most experimenters in any line had been able to do was to specialize on one single branch and die leaving the work unfinished?"[2] But he was confident, with a young man's confidence, that he would outdo his predecessors.

Naturally his brothers gave what support and assistance they could. The ten Burbank potatoes Gregory had allowed him to keep, which Luther regarded as his greatest tangible asset, were planted on George's land at Tomales. The first season's crop was carefully saved and replanted, and by the end of the second season, the yield was great enough for the potato to be put on sale. He ran an advertisement in one of the local papers:

BURBANK'S SEEDLING

This already famous Potato is now for the first time offered by the originator for trial on this Coast. For description see *American Agriculturist*, for March, 1878. PRICES: 1 lb. by mail, 50 cts.; 3 lbs. by mail, $1.00; 25 lbs. by express, $5.00.

LUTHER BURBANK, Nurseryman.
Santa Rosa, Sonoma County, California

This supplemented his income a little, but he had to contend with a certain amount of consumer resistance. People in California were used to red potatoes, and it took a while to convince them that a white one, though larger, smoother, and more productive, was an improvement. Later Burbank's Seedling became the chief potato grown on the Pacific Coast—by which time, however, Luther himself had long since ceased to grow it. He had no patent. It was only in those first few years before it became widely cultivated that it was worth his while to raise it for seed.

In the summer of 1877, worried perhaps by news of his illness, his mother and sister had followed him to California. Olive bought a house at the corner of Tupper and E streets in Santa Rosa, with four acres of land adjacent, part of which, Emma says, "Luther immediately rented and began his nursery. This land he later purchased, building a greenhouse thereon."

During the day, he worked as a carpenter. In the evenings, he took care of his growing plants and trees. He started with fruits and vegetables that seemed likely to sell locally. At that time, Santa Rosa was still largely wheat country. It was only subsequently that the local farmers began to think in terms of fruit growing. The market for seedlings was, therefore, an uncertain one. And, he remarks, even had it been more certain, "it would doubtless have been difficult for me to get a start, because fruit trees cannot be brought to a condition of bearing, or even to a stage where cions for graftings are available, in a few weeks. And I had neither capital nor credit, being virtually a stranger in a strange land."

But he had a good local reputation as a carpenter by now, and whatever he could earn with his hammer, over and above the barest necessities, went back into stock for his nursery business. "He was also employed by several American and European seed firms as collector of seeds of California's native plants," Emma tells us. "This gave him a very delightful way to become further acquainted with the trees and plants growing in this state. The knowledge thus gained of the locality, time of blooming and seed ripening, and other particulars, afterwards was of great value in his work."

In many ways, the time was ripe as far as the nursery was concerned. The first shipment of California fruit (it consisted entirely of Bartlett pears) had left Sacramento for New York within a year of the completion of the Central Pacific line. The consigning of California produce to eastern markets would soon be facilitated by the introduction of refrigerated cars. Already, Howard observes, "daring spirits were shipping fruit to faraway Chicago and receiving fancy prices. . . . The times were favorable for starting a nursery as existing concerns could not keep up with the demands for planting-stock. Especially was this true of prunes because that fruit could be sun-dried and sent to distant places with little danger of injury in transit. This looked like a safe crop and plantings, accordingly, were large."[3]

According to Burbank's old account book, sales for "nursery stock and ornamental and flowering plants" totaled only a modest $15.20 in 1877, but since he had not begun planting until summer of that year, this was hardly surprising. The following year, he recorded sales amounting to $84 ($70, according to Emma, who was teaching in Santa Rosa public schools to supplement the family income), and by 1879 they had reached $353.28. In 1880 the earnings of the nursery had climbed to $702, and the year after that, they reached the impressive total for those days of $1,112.69. Thereafter, business improved by leaps and bounds. Within ten years, the quality of his trees and the general reliability of his stock had become so widely known that he was doing more than $16,000 worth of business annually.

Even in those early days, he showed the sort of initiative that was to put him ahead of all competitors. His first catalogue, a twelve-page brochure issued in 1880, lists a hundred species of seeds "all gathered in South and West Australia and New Zealand during the past season." How he obtained them is not clear. (He would later make good use of varieties sent him by foreign collectors and correspondents, some of whom he employed for this specific purpose, but that source of supply had yet to be developed.) In any case, it showed notable enterprise, foreshadowing by years the later work of the Division of Foreign Plant Introduction of the United States Department of Agriculture.

In March 1881, Warren Dutton, a prosperous Petaluma merchant and banker, who had had the opportunity of seeing some of Burbank's work, approached him with an order for 20,000 prune trees, ready to be set out that fall. Dutton had taken a sudden interest in prune growing as a commercial prospect, and wished to start out on a large scale as soon as possible.

Burbank's first thought was to say that no one on earth could undertake to supply such an order at such short notice. But after a few minutes' reflection, he decided that it might just be done if almond seedlings were used for stock and prune buds "June-budded" onto them.

Dutton agreed to provide financial support during the summer to pay for the almond seedlings and the additional help that would be needed. The project would depend on the fact that almonds, unlike most other stone fruits, sprout almost at once. June-budding, if it is to be successful, calls for a long growing season.

Burbank set to work immediately. He had two acres of land available at the nursery, and he found another five he could rent. Almonds were bought for the planting and spread on a well-drained bed of creek sand. They were covered with coarse burlap cloth, and another inch of sand was spread on top of this. The almonds could thus be examined when necessary simply by lifting up one end of the cloth.

In less than fourteen days, some of the seeds were sprouting. These were removed and planted in the nursery. The others were covered up again and set out four inches apart in rows four feet from one another, as fast as sprouting occurred. The ground was kept carefully cultivated. In the meantime Burbank arranged with a neighbor to get a supply of 20,000 prune buds. By the end of June, the almond seedlings were large enough to be budded, and in July and early August, he labored with a large force of skilled help to bud the prune buds into them. After about ten days, the buds had made good unions with the stalks, and the tops of the little trees were broken and left hanging—in this way forcing the growth into the bud without endangering the trees themselves, which might have been killed if the almond twigs and leaves had simply been cut off. When the prune buds began to grow, they were tied up along the stalk, and when they were a foot or more high, the old almond tops were removed. By December 1, 19,500 trees were ready. The remainder could be supplied the next season.

Dutton was naturally delighted. He had been assured by other nurserymen that it would be impossible to produce the trees in the allotted time. Burbank, he let it be known, was a "wizard"—probably the first time this fateful title was applied to him.

"Although this was a horticultural stunt that had been employed by others," Howard remarks, "Burbank deserves full credit for having had the enterprise and initiative to do something that was new to his locality if not to the state. In the southern states the process is known as June budding or force budding, and is widely used in the propagation of peaches."[4] This achievement was the beginning of Burbank's reputation as a producer of prune stock. Other purchasers, who bought prunes in smaller lots, were equally satisfied. He took great pains to ensure that no tree left the nursery unless it was exactly what it was represented to be. The prune industry was expanding, and sometimes there was quite a crowd of would-be purchasers at his door, waiting their turn. It became something of a local jest in Santa Rosa when someone was being looked for, to say, "Well, if you can't find him in town, he's probably out at Burbank's Nursery waiting for some trees."

By 1884 Burbank was well established in the nursery business, with an income from it in excess of $10,000 a year. "Nothing more was required," he observes, "than to continue along the lines of my established work to insure a life of relative ease and financial prosperity." But in fact he had just reached the point at which he could afford to turn his attention to his true lifework. The Burbank story had only just begun. His ambitions went beyond being a mere nurseryman. The ferment that had begun with Darwin and the talk of his cousin Levi, the pupil of the great Agassiz, was coming to a head. "So," he says, "from the very hour when my nursery business had come to be fully established I began laying plans for giving it up."

Nursery and market garden had been a necessary apprenticeship. He had learned a great deal about the art of growing plants that was basic to his later work. At the same time, he had come to know the native California plants. In 1880 and 1881, he had made trips to the Northern California geyser country, where he found promising new material. Everywhere he went, he carefully studied the vegetation with an eye to its possible use in his future work. In the years to come, this knowledge would suggest novel ingredients for his hybridization experiments.

He would have liked to have carried these botanizing expeditions farther afield, and after 1884 it would have been perfectly feasible to have done so. The work of the nursery could have been left to assistants, while Burbank himself traveled the world in search of plants that could profitably be imported. But at heart he was an experimenter rather than a collector, and he was too eager to begin work to allow himself the luxury of travel.

He had already ordered seeds and cuttings of "a great variety of fruits" from Japan. In anticipation of their arrival he had bought the "Dimmick place" on Santa Rosa Avenue, four acres of neglected, run-down land that had been on the market for a number of years. This property, which was to be known as the Experimental Gardens, he immediately set about improving. Tiles were laid for drainage, and his neighbors marveled as 1,800 loads of manure were brought in. The first batch of Japanese seeds and seedlings arrived on November 5, 1884. "I felt," he says, "that a new era had begun for me."

While browsing in the Mechanics Library in San Francisco in the early eighties, Burbank had come across a book describing the travels of an American sailor, who in the course of a visit to the province of Satsuma in Japan had eaten plums with deep red flesh. This prompted him to order a shipment of seedling plum trees from Isaac Bunting, an export-import agent in Yokohama. Those that reached him in November 1894 had, alas, failed to survive the journey as a result of not being properly cared for in transit. Burbank wasted no time in reordering, and a second consignment arrived the following year, on December 20, 1885. This time the trees were in good condition.

There is some question as to exactly what Bunting sent. According to Howard, a memorandum found among Burbank's papers after his death listed twelve varieties of plum in all: one hundred specimens of the Blood Plum of Satsuma and ten each of the others. Burbank's editor and ghost-writer Henry Smith Williams makes him refer to only twelve seedlings. This discrepancy Howard attributes to Williams's desire to "glorify Burbank and make a great hero of him." A professional popularizer, he tended to make the story as dramatic as possible, without too much regard for accuracy.

California is the leading plum-producing state in the country, accounting for about 90 percent of the U.S. output of commercially grown plums and prunes. ("Plum" and "prune" were originally synonyms, used to describe hundreds of varieties of fruit falling into at least fifteen different species, but the latter term has come to designate those kinds that lend themselves to drying without removal of the pit and refers both to the fresh fruit and to the dried form.) Even before the planting of plums in the gardens of the Spanish missions, the California Indians made use of the fruit of the wild native species. The early settlers made some effort to cultivate these wild plums, the Pacific, Western, or Sierra plum, and the Gray Branch or Sisson plum, and one observer noted in 1858 that this indigenous fruit was potentially "a worthy competitor for flavor among our best varieties of gages and damsons." As the Spanish settlement proceeded, European plums were imported, and on his visit to California in 1792, George Vancouver came across them at Mission Ventura. The Mission prune, a variety about which little is known, is said to have been grown at Santa Clara as recently as 1870.

Plum varieties were brought to Sacramento from Oregon in the spring of 1851 by Seth Lewelling. That same year, a Mr. Shelton of San Francisco is recorded as having imported six varieties from Valparaiso, Chile, and by 1853 a Sacramento nursery, Warren and Sons, was listing eighteen kinds of plum. Another Sacramento nurseryman, A. P. Smith, who started business on the American River as early as 1848, listed twenty-eight plum varieties in his catalogue in 1856. In the early days, California growers were largely stocked by eastern nurserymen, but shipping by boat around Cape Horn was expensive, and the trees did not always arrive in the best condition. California nurseries soon sprang up to supply the demand. A census of 1859 found a total of 105,631 plum trees in the state. Plums sold for about fifty cents a pound in Sacramento at that time. Nursery trees were sold for as much as six dollars apiece in 1856, but in 1858 and 1859 Northern California nurserymen combined to fix a code of fair prices. A price of fifty cents to a dollar was established for year-old plum trees, one to two dollars for two-year-old trees.

The U.S. Patent Office is said to have imported the French prune in 1854, but California prune production is dated to December 1856, when Pierre Pellier brought scions of the French and Gros (Pond) prunes to San Jose. Some of these were top-grafted onto orchard trees by George W. Tarleton and J. Q. A. Ballou, and in 1859 Ballou sent the first shipment of California-grown French prunes to market in San Francisco. Some years later, in 1867, he was also the first to ship prunes around the Horn. By 1870 there were more than 19,000 prune trees in the state.

Prior to 1870, the only plums grown in California, if we exclude the wild species, were European varieties. In that year, however, a Vacaville man named Hough imported the Kelsey plum from Japan, through the U.S. consul there, paying ten dollars a tree. His stock was bought by John Kelsey, a Berkeley nurseryman, who subsequently contributed his name to this variety, also known as the Botankin. Another Japanese plum, the Chabot, was imported in the late 1870s by M. A. Chabot of Oakland, along with Japanese tea and bamboo plants. Burbank later acquired this variety from Chabot and put it on the market. There were, therefore, at least two varieties of Japanese plum growing in California when Burbank began to import them. (Japanese plums are, generally speaking, larger than the European varieties and crimson or red rather than blue or purple.) The Kelsey remains important among the late varieties grown today.

In his nursery catalogue for 1887, Burbank offered the new Japanese plums for sale, along with numerous other items. Howard comments:

> The contents of the catalog showed that he was going in heavily for new things—novelties he called them—and although he did not claim to be the originator of these unusual things, he left that impression, a trick that was not unusual among nurserymen. Several promising seedlings were listed, among these were apples, pears, plums, peaches, figs, persimmons, olives, and oranges. He refers to himself, about this time, as a dealer in novelties. But while most nurserymen were con-

tent to let others find new and promising things as chance seedlings, he was importing from Japan things that for the most part were entirely new.

Among the plums listed were the Botan, the Chabot, the Long Fruit, the Masu, or Large Fruit, the Botankio, the Botankio No. 2, and the Blood Plum of Satsuma. The announcement was relatively modest. "He had not yet learned to make the fullest use of the printed word," Howard observes.

> It is true that he did expand slightly in describing the Blood Plum of Satsuma by stating that he had the only tree growing in America and that it had cost him $40 in Japan, but the statement was made in 8-point type! The price quoted was one dollar per tree or seventy-five cents each for dormant buds. The only explanation I can offer is that he did not realize what he had. This mistake was remedied to some extent in the supplement that followed the catalog, but even here his walnuts and chestnuts from Japan were given more publicity space than the plums. As it turned out, the walnuts are merely a curiosity and the chestnut only of minor importance, while the plums laid the foundation of a huge industry, particularly in California and South Africa, and are of considerable importance in many other states and countries.[5]

Just before his death in 1932, Leonard Coates, a veteran California nurseryman who had himself introduced a number of plum varieties and originated a prune called the Coates 1418, informed Howard that J. E. Amoore, a tea buyer in Japan, had come to San Francisco in the eighties and set up a company that imported thousands of plums under Japanese names—one of which, he thought, was probably the Satsuma. Howard suggests that Coates, even if he did have some basis to his claim, should not be taken too seriously since "they were professional rivals and business competitors. Coates had the reputation of being a man of high and honorable principles but in this case he seems to have been obsessed by envy or jealousy." Admittedly, Burbank did obtain some trees from other importers—among them Chabot. Burbank himself says:

> The Satsuma and Burbank were the only two among my 12 seedlings that were directly introduced, although sundry of the others subsequently had a share in the production of hybrid races. It should be

> recalled also that I had somewhat earlier introduced three plums of Oriental origin—the Abundance, Chabot, and Berkmans, that were also a direct product of Oriental stock grown and fruited by me *from seedlings purchased from other importers.*[6]

So many synonyms have been used at one time or another that it is virtually impossible to trace the provenance of some of these plums. The variety called Abundance, for example, seems previously to have been known as Botan, Botankio, Chase, Yellow-fleshed Botan, and Yellow Japan. The Berkmans, or Berckmans, was variously also called Botan, Sweet Botan, and White-fleshed Botan. To make matters even more complicated, another variety, the Willard, also went by the name Botan. In any event, it seems plain that the two important varieties received by Burbank in this batch were the Satsuma and the Burbank, the latter named after its importer at the suggestion of H. E. Van Deman, pomologist of the Department of Agriculture. In a letter to Van Deman, dated August 22, 1888, Burbank wrote:

> The yellow fleshed plum with which you were pleased is, in my opinion, *far superior*, taking everything into consideration, to *any other* of the Japan plums. . . . I have and shall call it Burbank (by your suggestion) as nothing like it has been seen here and no Japanese importer knows it. It and the Satsuma were imported by myself with 10 other kinds, arrived here Dec. 20, 1885. As in all Japanese trees this was mixed in a lot of inferior plums. In the 12 varieties (received under number and Japanese names only) I obtained 20 or more kinds. The first Satsuma was obtained by me by *sending a man* from Yokohama to Satsuma for the original tree which cost me $50.00 and $55.00 delivered in Santa Rosa. All the Japan Importers of San Francisco confess they do not know either of them, that they never saw or heard of them. Also that they are delicious.[7]

The Burbank proved to be enormously productive and was particularly successful in home orchards throughout the country. Not a plum of the highest quality, it nonetheless raised the standard in many areas where only native varieties could previously be grown.

It seems clear from Burbank's own statement that Abundance, another variety that became very popular, was not one of the seedlings imported directly from Japan by him, though Howard identifies it as such. According to Liberty Hyde Bailey, the doyen of American horticulture, it was the same as a variety called Wassu, introduced by J. L. Normand, but H. M. Butterfield of the University of California's College of Agriculture identifies the Wassu with the Burbank![8] It is to be hoped that we have not here added to the confusion: the present author denies any special competence in the history of plums and refers future investigators back to the sources.

A little more than a week after the arrival of the second, successful shipment from Yokohama, Burbank negotiated the purchase of the Gold Ridge farm at Sebastopol, in the hills near Santa Rosa. Acquired on December 28, 1885, it was intended solely for experiment. "With the development of the Sebastopol place," Burbank says, "a new phase of lifework began."

The original farm consisted of about ten acres of sandy soil some seven miles from Santa Rosa. Burbank purchased an additional five acres on one side and three on another, making a total of eighteen. "At the time the place was purchased," he says, "about two-thirds of it was covered with white and tan oaks, the native Douglas spruce, manzanita, cascara sagrada, hazel and madrona, while beneath the trees grew brodiaeas, calochortus, cynoglossum, wild peas, fritillarias, orchids, sisyrinchiums—yellow and blue—and numerous other wild plants and shrubs."[9] The land had the advantage of being well watered, and after it was cleared, numerous species of clover made their appearance. Turning these under in the spring provided an excellent source of nitrogen for the soil. Burbank was always very much aware of the importance of good soil culture.

In 1888 he sold part of his nursery business to R. W. Bell, a partner he had taken in. Bell took over the growing and marketing of the standard varieties, along with the old nursery buildings and about an acre of the old sales ground, while Burbank kept control of the novelties he had developed. A miscellaneous announcement of 1888, in the form of a postcard, introduced Bell as "my successor in the standard fruit and shade tree business. . . . I shall confine my business in future to olive and nut trees and horticultural novelties exclusively." To prevent confusion, Burbank retained the name "Santa Rosa Nurseries," whereas Bell's establishment was to be called "Bell's Nurseries."

In September of that year, divested of what was perhaps the

more troublesome side of his business, Burbank allowed himself the luxury of a trip back to Massachusetts. Though he was now a financial success, he still did not indulge himself by taking a sleeping car. The Puritan instincts died hard. He slept curled up on the seat "like a jackknife." A $1.50 comforter helped.

He reached Lunenburg on the stage from Fitchburg on the evening of September 19 and was immediately in the thick of his old friends. "You cannot imagine the handsqueezings and questions that I was put to for about eight hours," he wrote. "One of my old Lancaster Academy schoolmates (Nellie Day) said she waited hours before she could get an opportunity to speak." This was at the Fitchburg Fair, where he was among the speakers.

"The old place looks as it did," he reported. "Mr. Gregory expects to have a monument raised on the spot where the Burbank potato originated, so I pointed it out." Meanwhile, he gathered seeds and plants to take back to California and avoided the usual importunities of an expatriate's visit home. "Fanny Peabody says she has a kitten for Alfred, but I will be darned if I will carry a circus back. She will have to send it by mail."

He stayed scarcely two weeks. By October 3, he had said his good-byes, and was sight-seeing in Boston. From there he went on to Washington, where he arrived on the ninth, on what seems to have been an organized tour. "Saw the original Declaration of Independence, visited the White House, the President's reception room, Weather Bureau, Senate while in session, House of Representatives, Supreme Court (which is trying the Bell Telephone case)." There were eighty in his party. They were put up at the Willard Hotel, which cost $4.50 a day, and taken by steamer to Mount Vernon. There was even a presidential reception hosted by Grover Cleveland. "As his wife is off visiting, we shall not see her," Burbank observed. He had acquired "a great quantity of very rare wild and cultivated seeds" and paid a visit to the Department of Agriculture before he left.

The party returned to Boston on October 15, and a few days later he was in Lancaster again. He was, he says, "healthier, happier, and handsomer than ever you saw me; and if business would live through it, in any way, I would extend my time, for I have got more real happiness out of the last month than for fifteen years." All the same, he confessed, "I should never be contented to live here and no sensible Californian would either." By comparison with Santa Rosa, he found New England weather distinctly disagreeable and remarked that he would "probably have to buy a suit of oil clothes, rubber boots, and a boat, and some lightning rods, etc." An old friend asked him, "If you had a farm here what would you raise?" Luther replied, "I think I would try to raise enough money to get to California."

By early November, he was home in Santa Rosa again.

8

"NEW CREATIONS"

His holiday had one fateful result. On the train (it was probably on the way back to California), he had met a schoolteacher named Helen A. Coleman.[1] She is described as being a "youngish widow" from Denver, Colorado.

Long after his death, a reputation as an "adventuress" and a vamp would be spoken of. Burbank—they said—had had little experience of women and was an easy mark. What transpired between the couple in transit there is now unfortunately no way of telling, but it is true that she presently followed him to Santa Rosa—and stayed put there until they became formally engaged.

The headlong passion that Walter Howard suggests is probably to be doubted, for Burbank cannot be charged with undue haste. (The notion that he was grievously smitten seems to rest largely on his "extravagance" in buying a horse and phaeton for his fiancée's entertainment. But surely this was hardly out of the way for a man with an annual income now in excess of $10,000 and the prospect of doing better every year. In 1889 his contemporary William Allen White found a salary of eighteen dollars a week "beyond the dreams of avarice.")[2] More likely, Burbank simply felt that it was time he had a wife and children of his own. Himself one of fifteen, he was no stranger to family life. His love for children was widely publicized. He must doubtless have wanted a son, or sons, who would one day be able to carry on his work.

Smitten or not, horse and carriage notwithstanding, the couple waited nearly two years before going to Denver to be married. Only then did Burbank feel sufficiently sure of his feelings to take the plunge. Evidently he had doubts, and in the outcome they were fully justified.

In Denver Burbank "entertained her royally for about two weeks," Howard says, implying evidence of infatuation. The wedding itself took place on September 23, 1890. He was forty-one ("a highly susceptible age for a man," Howard remarks) and well on his way to being a national figure. Of Helen Coleman we know very little.

"She was peculiar in her dress and speech and even at home her manner was silly, simpering, and affected," informants who had known her in Santa Rosa later suggested. But that was almost half a century after the event, and such sources are notoriously unreliable. Judge Rolfe L. Thompson of Sacramento allowed that she had had some good traits of character. She was something of an artist, and he remarked that his family still possessed a picture she had painted. She was interested in literature, he remembered—a great reader.

In the squabbling that ensued, it would be only reasonable to suppose that, as is usually the case, there was a degree of justice on both sides. Helen had perhaps expected something very different. Denver, her hometown, was a rip-roaring metropolis. Santa Rosa was a small country place, with little excitement to speak of, once the initial romance had worn off. And Luther himself was hardly as gregarious as a wife might reasonably have hoped. If, as we may surmise, her ambitions had been kindled by meeting and marrying a dynamic and exuberant Burbank, newly returned from a triumphal tour of the native places that had once doubted whether he would ever amount to a hill of beans, as one local put it, she was probably in no proper frame of mind to settle down and keep house for a reclusive horticulturist.

In the end, there was really little in common between them. He cared only for his work, and she took no interest in that. She was "high-strung," as they used to say. She had a nervous temperament. From the beginning, trouble was inevitable with Luther's mother. Old Mrs. Burbank was almost eighty, but she was also of a very positive disposition, set in her ways and used to getting them. She had occupied the position of head of the family for a long time, both before and after coming to California. She was not about to surrender it to the intruder.

Emma may have been another problem. She, too, had lived with Burbank the greater part of his life. She had an almost absurd faith in his abilities. "She worshipped her brother as a divine dispensation," wrote George Shull, as reliable an observer as any. "She was very religious, and was quite convinced that God had specifically sent her brother to do His appointed miraculous work. With this viewpoint you can understand her solicitude that her brother should not be unduly imposed upon. He must be guarded and protected in every way possible so that he could achieve the specific results he was divinely appointed to get for the welfare and redemption of mankind."[3] Emma sang his praises to whoever cared to listen, kept up his scrapbooks of clippings, and stood between him and the hero-worshiping public that might otherwise have taken up more time than he could afford. "A true friend and disciple, devoid of ulterior motives, she never let him down," says Howard. "Incessantly she fed his natural ego and cultivated the Messiahship idea."[4]

In 1938 Howard interviewed D. B. Anderson, a dentist practicing in Santa Rosa. Anderson had done secretarial and accounting work at the Burbank home during the evenings in the years 1887–89. He recalled Burbank telling him "that the world had experienced many Christs—at least thirteen; that they assumed different forms and might continue to arise from time to time, and hinted that he, himself, he felt, was approaching that status." On another occasion, he had compared himself to Napoleon, saying: "See, I am about the same height as Napoleon and my hat is about the same size as his, although my head is growing and increasing in size all the time."[5] Anderson was of the opinion that these remarks were made in all seriousness, and although they bear the stamp of Burbank's whimsical sense of humor, it is very possible that they were nonetheless at least half-serious, as whimsy often is. Although admitting that Helen was cantankerous and led Burbank a dog's life from the first, Anderson did not go along with the general belief that she was crazy: she was, he said, merely trying to get rid of all his relatives.

It is hardly difficult, in this context of a vain but retiring husband, a possessive mother and sister-in-law living in the same small house, and friends and neighbors little appreciative of her Denver clothes and manners, to envisage the conflicts that must have lain in wait for the new Mrs. Burbank. And underlying all that, who knows what else? There were, in the upshot, no children.

Helen indulged herself in fits of temper of which Burbank bore the brunt. They were both very unhappy. He loved children and had encouraged them to play in his yard. Now she humiliated him by loudly ordering them off. When he came in from his work, she began to nag and continued until after they had gone to bed. She would not let him sleep, he later testified. Once she had tried to push him out of bed, and when she found she couldn't, got up, took hold of a revolver, and threatened to shoot him. Slamming a screen door in his face one day, she blacked both his eyes.

She declared that his version of this last incident was greatly exaggerated and that she had by accident let the screen door in the kitchen loose, not knowing that he was about to pass through. But he, for his part, felt so humiliated by his black eyes that he moved into a workroom above the stable, where he could lock himself in and have some peace. There he remained for two years.

And these were years in which his work load was particularly heavy. It was around this time that he first began to attract world attention (particularly after the appearance of his *New Creations* catalogues of 1893 and 1894). In a one-page announcement in 1892, he observed of his varieties: "My time is so wholly occupied in their production that I cannot attend to their introduction and will sell the stock and complete control of some of the most promising new fruits and plants, or would be a member of a joint stock company, for their introduction. Correspondence on the subject solicited." There seem to have been no satisfactory offers, and he was obliged to soldier on alone. There were hundreds of experiments under way on his grounds, needing daily attention and observation. It was always his tendency to take on more than he could handle. Perhaps he was glad to keep busy, to escape the unhappy atmosphere of his home. But he had never been robust, and work and wife combined were almost too much for him. "Under the double load," Howard says, "his health failed and he almost became a nervous wreck."

Ever since 1877, he had been working with walnuts, using more than a dozen species and varieties as breeding stock, among them the eastern and Californian black walnuts, the Persian or cultivated walnut (also called the English walnut, this variety is actually native to southeastern Europe and Asia), the butternut, and several Asiatic species, including the Japanese walnut, or heartnut, *Juglans sieboldiana*.

His greatest success was with the northern variety of the California black walnut, *Juglans hindsii*. Crossed with *Juglans regia*, the Persian walnut, and with the black walnut of the eastern United States, *Juglans nigra*, this gave rise to the "Paradox" and "Royal" walnuts respectively. Another noteworthy Burbank walnut was the "Santa Rosa Soft Shell," a cross between two Persian varieties, introduced in 1906.

In the Royal and Paradox* walnut trees, Burbank felt he had a major triumph. Both were notable for their size and rapid growth, and although the Royal also produced great quantities of nuts—as much as a ton from a single tree in one season—their greatest potential seemed to be as lumber. The Paradox, resulting from a wider cross than the Royal, showed the partial sterility found in many interspecific hybrids. It produced very few nuts, and though these readily germinated on planting, the second-generation seedlings displayed an extraordinary degree of variation. "In the same row there will be bush-like walnuts from six to eighteen inches in height side by side with trees that have shot up to eighteen or twenty feet; all of the same age and grown from seeds gathered from a single tree. This rate of growth continues throughout life, and the fraternity of dwarfs and giants has been a puzzle to layman and botanist alike," Burbank said. This was a major disadvantage, as unlike most fruit trees, walnuts cannot easily be propagated from cuttings.

* The introduction of this hybrid, under the name "Vilmorin," is sometimes also attributed to Felix Gillet, a pioneer nurseryman of Nevada City, California.

The increased capacity for growth in cross-bred individuals, both in plants and in animals, had long been known. This phenomenon is called "hybrid vigor," or heterosis (the latter term, introduced by George Shull in 1914, has now become general). It is exemplified in that best known of F_1 hybrids,* the mule, which, though sterile, can, as they used to say, "get along with less food than a donkey, do more work than a steam engine, and has more horse-sense than a horse." Sterility in reproduction, which may be partial, as in the case of Burbank's Paradox walnut, or complete, as in the mule, is the first indication that the limits of hybridization are being reached. Even where the hybrid is fertile, subsequent generations may show a failure to breed true, and there may be a rapid falling off in vigor. "The mule would quickly forfeit its place in the estimation of teamsters could it bring forth the mongrel progeny which would certainly ensue if it were fertile," remarks the geneticist Donald Jones. "Animal husbandry men have learned by sad experience that those cross-bred cattle, sheep and swine which are so profitable to feed for the market are of no value for breeding purposes." Burbank's second-generation seedlings could not, he points out, be expected to equal the first-generation hybrid trees that he described in such glowing terms and upon which the calculation of probable returns was based.[6]

It would be many years before these processes were understood, however, and Burbank's attention was focused on his amazingly vigorous first-generation hybrids. The Paradox walnut, he declared, "so far outstripped all competitors in the matter of growth that it might fairly be said to represent a new type of vegetation." At sixteen years of age, the trees stood sixty feet high, with an equivalent span in branches. Four feet above the ground, their trunks had a diameter of two feet. Thirty-two-year-old Persian walnuts growing nearby averaged only eight or nine inches in diameter and had a spread of branches only about a quarter that of the hybrids. Moreover, the wood of the Paradox walnut was harder and more close-grained than that of the ordinary black walnut. This was remarkable inasmuch as trees that grow rapidly usually have soft wood. "All in all the production of the Paradox hybrid, and the development of a race of hard-wood trees of ex-

* In this notation F_1 means first filial generation, F_2 second filial generation, and so on.

ceedingly rapid growth, constitutes a genuine triumph in tree culture," Burbank pronounced. "A tree that grows to the proportions of a handsome shade tree and furnishes material for the cabinet-maker in six or eight years, has very obvious economic importance."[7]

The Royal walnut, like the Paradox, grew more rapidly than its parent varieties. Not being the product of a cross between such distantly related types, the second generation did not show such a wide range of diversity, and the Royal was also extraordinarily fecund. "The first generation hybrids," Burbank said, "probably produce more nuts than any other tree hitherto known." He conceded that because of the tendency to "throw back" to the grandparent strains, seedlings grown from the nuts could not be relied on to reproduce all the good qualities of their parents. Propagation was therefore best accomplished by grafting, and he suggested that Persian walnuts be grafted on Royal root stock, arguing that "a tree grafted on this hybrid will produce several times as many nuts as a tree of corresponding size growing on its own roots. The trees are also much less subject to blight when they are thus grafted." This would, however, be a relatively expensive process if the trees were to be grown solely for lumber, and the nuts, like the original black walnuts, were thick-shelled and difficult to crack. "The slimy deep-staining hull is not easily removed," notes Donald Jones.

Both the Royal and Paradox walnuts were offered for sale (though not yet under those names) in the first edition of what was to be Burbank's most important single catalogue, *New Creations in Fruits and Flowers*, issued in June 1893 and bearing the bold subscription, "*Keep This Catalogue for Reference*. You will need it when these Fruits and Flowers become standards of excellence."

This fifty-two-page brochure was the means by which Burbank first became internationally famous (he was already well-known in the United States) and announced some of the best and most lasting things that he was ever to produce. It went out from "Burbank's Experiment Grounds"—with "great care being taken to confine it to the trade only," according to a note in a subsequent catalogue. Burbank wanted it to be clearly understood that he was an experimenter, an innovator, and not just another nurseryman.

An announcement drove this message home with unusual forcefulness, and what to some might have seemed no small degree of arrogance:

> The Fruits and Flowers mentioned in this list, and to be mentioned in succeeding lists, are more than new in the sense in which the word is generally used; they are new creations, lately produced by scientific combinations of nature's forces, guided by long, carefully conducted, and very expensive biological study. Let not those who read suppose that they were born without labor; they are not foundlings, but are exemplifications of the knowledge that the life-forces of plants may be combined and guided to produce results not imagined by horticulturists who have given the matter little thought.
>
> Limitations once thought to be real have proved to be only apparent barriers; and as in any of the dark problems of nature, the mental light of many ardent, persevering, faithful workers will make the old paths clear, and boundless new ones will appear by which the life-forces are guided into endless useful and beautiful forms.
>
> We are now standing just at the gateway of scientific horticulture, only having taken a few steps in the measureless fields which will stretch out as we advance into the golden sunshine of a more complete knowledge of the forces which are to unfold all the graceful forms of garden beauty, and wealth of fruit and flowers, for the comfort and happiness of Earth's teeming millions.

This catalogue, Burbank advised his public, gave the results of more than twenty years' work: these were "the *best of millions* of cross-bred, hybrid and seedling plants, which are now and have been produced at the rate of a million or more a year." Besides the new walnuts, *New Creations* offered a "New Japan Mammoth Chestnut"—to produce which, Burbank said, he had for many years grown and fruited seedlings by the thousand. There were two new quinces, the "Van Deman" and the "Santa Rosa," each offered for $800. The Van Deman was named, with consent, for Professor H. E. Van Deman, head of the Pomological Divison of the Department of Agriculture; along with other Burbank seedlings, this quince had received the Wilder Medal from the American Pomological Society in 1891. The Santa Rosa came with a testimonial from W. A. Taylor, assistant pomologist with the USDA, to the effect that it was "very choice" and could be eaten raw like an apple. There were, besides, a new Japan quince, the "Alpha," and a new flowering quince called "Dazzle," each offered for $300.

Among the most spectacular offerings were his new hybrid plums and prunes, ten of them in all. Two crosses between the Petite d'Agen, or French prune, designated "A.P.–90" and "A.P.–318," were offered for $1,500 and $3,000 respectively. This was for stock on hand and complete control. A.P.–318, which, Burbank contended, might "change the whole Prune industry of the world," could also be had in half-shares: he would sell half his stock and half control for $2,000. The "Golden" plum, a cross between the Robinson Chickasaw and the Sweet Botan, of which Burbank noted, "I have never seen a Plum tree which perfects so much fruit, and has it so evenly distributed as this one," was priced at $3,000. Stock and control in another hybrid plum, "Perfection," a cross between the Kelsey and the Burbank, were offered for $2,000. "Among the many thousand Japan Plums which I have fruited, this one, so far, stands preeminent in its rare combination of good qualities," Burbank said. He was not wrong. Renamed the Wickson, it was to prove one of his most enduring varieties. Six other hybrid plums, most of them with ancestors in the original Yokohama shipment, were listed at prices ranging from $300 to $500. Rarely in horticultural history can so many important new varieties of plum have been announced in one catalogue.

Burbank was particularly proud of his work with berries. Nothing like it, he said, "has anywhere ever been attempted even by Government aid; and no one will question the claim that I have made more and greater improvements in Blackberries and Raspberries during the last fourteen years than have otherwise been made during all the past eighteen centuries."

He goes on to list thirty-seven species of the genus *Rubus* that had gone into the production of his hybrids, remarking that "the combinations are endless; the results are startling and as surprising to myself as they will be to others when known." Among those on offer was the Japanese golden mayberry, or "Improved Rubus palmatus," with a list price of $800. This Burbank claimed to be the earliest-bearing raspberry ever known. He gives a short account of its history, which is highly descriptive of his methods:

> Some ten years ago I requested my collector in Japan to hunt up the best wild Raspberries, Blackberries and Strawberries that could be found. Several curious species were received the next season, and among them a red and also a dingy yellow unproductive variety of

> Rubus palmatus (described by botanists and collectors as being unproductive and having an insignificant berry). One of these plants, though bearing only a few of the most worthless, tasteless, dingy yellow berries I have ever seen, was selected solely on account of its unusual earliness, to cross with Cuthbert and other well-known Raspberries. Among the seedlings raised from this plant was this one, and, though no signs of the Cuthbert appear, yet it can hardly be doubted that Cuthbert pollen has affected some of the wonderful improvements to be seen in this new variety.

Another berry of great interest, offered for $600, was the "Primus." This was a cross between the hardy Siberian raspberry, *Rubus crataegifolius*, and the cultivated variety of the dioecious* California dewberry, *Rubus vitifolius*, known as Aughinbaugh. Five hundred seedlings were raised, but only a few showed any promise. One of these, a trailing plant showing many of the properties of both parents, was selected. The fruit, which was of a dark, mulberry color, was something like a blackberry in form. "The flavor is unique; nothing like it has ever before appeared; all pronounce it superior when cooked, and most people like it raw, while some claim it is the 'best berry they ever tasted,'" the catalogue declares. The most remarkable feature about this cross was, however, that *the progeny came true from seed*. Seedlings closely resembled the parental bush. Jones has speculated that what Burbank had here is what afterward came to be known as an amphidiploid or allotetraploid,† a species hybrid that bred true because of an increase in the number of chromosome sets. If so—though Burbank could not have been aware of why his hybrids were fertile and bred true—this may be the first recorded report of this phenomenon. In the early years of genetics, when Mendel's rules of segregation and recombination were being tested and proven, scientists tended to discount such findings. Here, Jones observes,

> was a case where a plant hybrid bred true in the second generation where, by the rules, it should have shown the greatest breaking up into variable types, especially where the parental stalks were so diverse. Away with this imposter who claims such unorthodox results! Either the cross was never made in the first place or else the plants were mixed up with something else while being grown. Such results simply could not be.

* That is, with staminate and pistillate flowers borne on different individuals.

† See Appendix I.

Many years later, with Mendel's law carefully proven, biologists have gone further into the subject and have found many species hybrids: in tobacco, in cabbages, in tarweeds—strange plants from unholy marriages—that have given offspring that breed true right from the start. From Russia comes the weird report of a cabbage-radish hybrid that produces neither succulent head nor fleshy root and has flowers and seed pods unlike any brassica ever seen before. When it is examined under the microscope this plant is seen to have 36 chromosomes, those rodshaped carriers of the inheritance, instead of the 18 of the radish and cabbage. Abnormal chromosome distribution and duplication of chromosomes are topics of interest at the present time in biological laboratories devoted to the study of heredity and related subjects. This cabbage-radish plant also bred true from seed and comes the nearest to being a new species that has actually been produced under experimentally controlled conditions, unless the Primus berry that Burbank made before 1889 is a better example.[8]

More than a dozen other varieties of berry were offered for sale in the 1893 *New Creations*. Among these we might note two crosses between the improved California wild dewberry and the Cuthbert raspberry, listed at $800 each. These were larger than the largest berry ever before known, Burbank claimed, asserting that individual berries often measured three inches around one way by four the other. When they were exhibited, the question asked was, "Will they be sold by the dozen?" A hybrid raspberry designated "S.S.–147" was said to be "the first practical cross of the cap and sucker Raspberries ever made . . . no berry has ever been introduced which so delightfully combines the best flavors and aromas of both these species." This was listed at $400. A hybrid berry called "Paradox," which seems to have closely resembled the modern boysenberry, was priced at $800. This was said to have "appeared in the fourth generation from a cross of Crystal White Blackberry and Shaffer's Colossal Raspberry. The plant is in every respect a most perfect balance between the two species."

Aside from its cornucopia of new berries, plums, and prunes, *New Creations* goes on to list five new seedling roses; a number of new callas, with prices in the $1,000–$2,000 range; two new hybrid lilies; a new myrtle; a new poppy named "Silver Lining"; hybrid nicotianas; a "Begonia-leaved" squash described as "mammoth . . . produces abundant crops for stock-feeding"; three new varieties of potato; an ornamental tomato, "Combination"; and a new plant, the "Nicotunia," produced by crossing the large flowering nicotianas with petunias. Of this last Burbank says proudly:

> If one thinks he can take right hold and produce Nicotunias as he would hybrid Petunias or cross-bred Primroses, let him try; there is no patent on their manufacture; but if the five hundredth crossing succeeds, or even the five thousandth, under the best conditions obtainable, he will surely be very successful; I do not fear any immediate competition.

Though the Nicotunias produced no seed, they could easily be propagated from cuttings. The whole stock, however, was killed off by frost not long afterward.

The catalogue concludes with a list of other hybrids Burbank was experimenting with. They included crosses of peaches with almonds, plums, and apricots; of almonds with plums; of apricots with plums; of quinces with apples; and of potatoes with tomatoes. "Not only one of each of these is growing, but many, in some cases several hundreds each, and of some, thousands," he asserts.

Finally, at the end of the catalogue, there is a sort of Burbankian position paper entitled "Facts and Possibilities," which is worth quoting at some length, inasmuch as it is the clearest statement of Burbank's theories as of that time. It cannot be doubted, he says,

> that every form of plant life existing on the earth is now being and has always been modified, more or less, by its surroundings, and often rapidly and permanently changed, never to return to the same form. When man takes advantage of these facts, and changes all the conditions, giving abundance of room for expansion and growth, extra cultivation and a superabundance of the various chemical elements in the most assimilable form, with abundance of light and heat, great changes sooner or later occur according to the susceptibility of the subject; and when added to all these combined governing forces we employ the other potent forces of combination and selection of the best combinations, the power to improve our useful and ornamental plants is limitless. But in crossing, as in budding or grafting, the affinities can only be demonstrated by actual test, which often involves long, tedious, and expensive experimenting.
>
> In budding or grafting, the nurseryman finds every conceivable stage of congeniality between stock and bud or graft, from actual poisoning to a refusal to unite; or uniting and not growing; or growing for a short time and dying; or separating where united; or bearing

one or two crops of fruit and then suddenly blighting or separating after years of growth up to complete congeniality. So in crossing, all grades of hybridity are to be found. Crossed plants generally have the characteristics of both parents combined, yet, owing to prepotency of the life-forces in certain directions or congeniality of surroundings sometimes show only their parentage on one side producing uncertain results in the first generation, and these cross-bred seedlings often break away into endless forms and combinations, sometimes even reverting to some strange ancestral form which existed in the dim past; or the break may not occur until after many generations, but when once the old persistent type is broken up the road is open for improvement and advance in any useful direction. Sometimes hybridized or crossed seedlings show considerable or even great variations for weeks and then change at once to one or the other of the original types; or they may show no change in foliage or growth from one or the other parent forms until nearly ready to bloom or bear fruit, when they suddenly change in foliage, growth, character and general appearance.

Tomatoes may be grown from seed pollenated from Potato pollen only, and Juglans regia from nuts pollenated only from Juglans cinerea or J. nigra. The common Calla has often been grown from seeds pollenated only by Calla albo-maculata; also pure Wheat from Rye pollenations, and *vice versa*; pure Blackberries, Raspberries and Dewberries from Apple, Rose, Quince or Mountain Ash pollenations.

Seedling Lilies very rarely show the effect of foreign pollenation, though often producing seed much more abundantly than with pollen of the same species.

These facts have been observed by me so often, and have been worked on so extensively, and can be proven so readily, that the common theory of parthenogenesis must, in these cases, be set aside.

There is no barrier to obtaining fruits of any size, form or flavor desired, and none to producing plants and flowers of any form, color or fragrance; all that is needed is a knowledge to guide our efforts in the right direction, undeviating patience and cultivated eyes to detect variations of value.

The descriptions in this list are necessarily short and incomplete, but in all cases exaggeration has been studiously avoided.

It is fair to suppose that one who has had extensive experience in any special line should be able to give judgment approximating impartiality. All the new plants mentioned in this list and the supplementary lists which are to follow will have to be judged by the great discriminating public, and will infallibly stand or fall by its verdict, without regard to what the originator or introducer may see fit to say.

It should be remembered that this was written at a time when the results of hybridization could not be predicted. It was only with the rediscovery of Mendel's principles of heredity almost twenty years later that what was called "prepotency" was identified as Mendelian dominance and "reversions," or "atavism," as the reappearance of Mendelian recessives. Until that time, as Edwin Conklin has remarked: "The fact of evolution was accepted by practically all scientists, but the factors of evolution were largely matters of opinion, and in general persons believed what they preferred to believe. Indeed this whole subject had become so speculative that it seemed to be a field for the exercise of the imagination rather than of scientific research."[9]

New Creations is Burbank before the Burbankians—before the puffers and front-office men who are generally credited with having inflated his reputation out of all proportion to the truth. Yet already it makes amazing claims—some of them a little hard to believe even with the best will in the world. "Exaggeration has been studiously avoided," he insists. Yet we know that he was by nature given to exaggeration. (George Shull held that he had an "exaggeration coefficient" of about ten; that all his figures should be divided by this number to get an approximation of the truth.) As a nurseryman he almost had to exaggerate. It was considered fair practice in the trade, his competitors all did it, and if his varieties were to hold their own, they needed the customary boost. His experiments were entirely funded by himself. Their continuation was dependent on sales. If he could not make a healthy profit, he might as well shut up shop.

Burbank refers his introductions to the judgment of the "great discriminating public," and if we judge by the reception they got—at this late date there is hardly any other way—they must have had a substantial part of the merits he claimed for them. Though his prices were high, they sold, and went on selling.[10]

The biggest single purchaser was Stark Brothers' Nurseries of Missouri, which acquired the Golden plum for $3,000, the A.P.–318 prune for the same amount, the Van Deman quince for $800, and a half-interest in a hitherto uncatalogued plum, called the "Doris," for a probable $300. According to one version, Clarence Stark, the head of the firm, told Burbank: "I don't think you will ever make a real success in the nursery business because your heart is not in it. But if you will carry forward the type of hybridizing you are doing, I think you will go very far in your chosen field. To demonstrate our sincere belief in your work, our company will give you $9,000 if you will let me pick three of these new fruits you have shown me." The Van Deman quince continued to be listed in Stark Brothers' catalogue more than eighty years later.

"I had been getting very nominal and insufficient prices for my new fruits, and when Mr. Stark made this offer to me that is when I really became the Luther Burbank that horticulture has known," Burbank is said to have remarked. "I could see a vision of the great possibilities of plant improvement if only sufficient time, money and effort were put behind the work by those who understood it."[11]

In addition to the sales to Stark Brothers, Burbank received over $5,900 from John Lewis Childs of Floral Park, New York, and other eastern buyers also spent freely on his "new creations" of 1893. These were all experienced nurserymen. It is unlikely that they were motivated by altruism or would have been taken in by inflated claims. That they were not disappointed in their purchases is evident from the fact that a number of them remained steady customers in the years that followed.

9

BURBANK THE WIZARD

In the months that followed the publication of *New Creations*, orders came pouring in. The catalogue itself was in great demand. "Probably no horticultural publication ever created more profound surprise or received a more hearty welcome," Burbank wrote in the introduction to a new edition, *New Creations in Fruits and Flowers* of June 1894. "Almost every mail brings requests for them from colleges, experiment stations, libraries, students and scientific societies in Europe and America, and it has been translated into other languages for foreign lands, even where it would seem that scientific Horticulture was hardly recognized; some asking for one, others for two or three, or a dozen or two or more." Henceforth there would have to be a charge for the catalogue.

Letters came pouring in by the thousands, many of them from amateurs who had long lists of questions they wanted answered. Unless this avalanche slowed up, Burbank said, "there will soon be no one here to answer them." The sudden fame he was experiencing seems to have gone to his head, for he now adopts the royal plural, declaring, "We love to produce new fruits and flowers, and our heart is made glad beyond expression to know that our work is appreciated far and wide."

At the same time, he endeavors to respond to the criticisms that had begun to be heard. His products, he insists, will not be found any less hardy as a consequence of being raised in the mild California climate. "Are those already before the public any less hardy or any less valuable than most of the Russian fruits which have been so extensively advertised for years?" Nurserymen enterprising enough to have invested in his trees are reaping a rich reward. And the best are yet to come.

> We have the fullest sympathy with all nurserymen who desire to introduce only those trees, fruits and flowers which have been widely tested, and which will prove better than any before known; delicious, productive, handsome, hardy and reliable in every respect. *We would very much prefer to have all our new fruits and flowers fully tested everywhere* and by everybody; but those who know the facts are too well aware that it would be a perilous risk or utter ruin to the originator, as a single bud or seed in the wrong hands may place an unscrupulous person on an equal footing with the originator, who may have spent worlds of patient thought and toil, during the few short years of the best of his life, in producing the beautiful creation. Having no Government aid or even protection, or college endowment to back us and to pay our bills, we must receive early returns, in part at least, for our tremendous expenses. Most of the horticultural world now knows that we would not send out a plant which was thought to be unworthy; else would the keenest and most level-headed business men—men who have built up mammoth horticultural establishments which are a wonder far and near—pay thousands of dollars for a single plant or tree which they had never seen, to one living in a far-off State or nation whom they had never met?

The originator should not be blamed, he points out, for indiscriminate and unwarranted praise on the part of the purchaser. "Indiscriminate commendation has a tendency to discourage all the honest efforts of originators who might, perhaps, otherwise receive some reasonable compensation for their labors." This was certainly to be true in his own case. Some of the most severe criticism that was to be leveled at him had its origins in other nurserymen overpraising varieties that bore his name. On the other hand, he was too often guilty of professional exaggerations himself to expect much sympathy from his colleagues on this score. If anything, they had merely followed his lead.

The 1894 edition of *New Creations* offered a number of further varieties for sale and reiterated some of the announcements in the previous catalogue.[1] Prominent among the new offerings was the Wickson plum, named for Burbank's friend Edward Wickson of the University of California, author of a popular book on *California Fruits and How to Grow Them*. This cross-bred Japan plum was to be one of the most enduring of all Burbank's introductions. "A year ago I was convinced that this was perhaps the best of all the Japan Plums," he says, "and have yet no reason to change that opinion. . . . There is wood enough now for twenty thousand grafts or half a million buds; and if not sold before September 1st, shall introduce it myself to the general trade. Price, $2,500."

The Wickson was the "Perfection" plum of 1893, a cross between *P. triflora* and *P. simonii* that became the leading shipping plum for several years. "As the namesake of this plum, the writer has inherited both joy and embarrassment," Wickson himself later wrote. "To read in some eastern horticultural reports that 'Wickson is a worthless cross-bred Japanese' has caused him some anxiety lest he might be deported to Tokio."[2]

The charge was made that the Wickson and other Burbank varieties were simply Japanese plums brought to this country and renamed. But there is scant doubt that they were, in fact, the hybrids they were said to be. Evidence of this, Jones remarks,

> was sought from Mr. T. Tanikawa, pomologist in charge of the Horticultural Experiment Station at Okitsu in Japan. His reply was to the effect that the Wickson variety resembled most closely the native plums in tree, fruit and leaves but that he had never seen exactly the same variety in Japan. Most of the other Burbank plums of this type, he said, are quite similar to Japanese varieties in fruit but the tree and leaf characters are different.[3]

One of the most curious items in the catalogue was the "Iceberg" white blackberry. Burbank's attention had initially been attracted to the possibility of producing a white blackberry as a novelty by the Crystal White variety. This was not a true white blackberry as the fruit was more of a brownish yellow when ripe. Burbank's approach to the problem was entirely experimental. He crossed the Crystal White with the Lawton, a black variety. In the first (or F_1) generation of offspring, the seedlings produced only black fruit, but when they were interbred, a few plants with light-colored fruit turned up (F_2). The seed from these was planted, and in the next (F_3) generation, four or five bushes were found, out of a total of several hundred raised, that had almost completely colorless fruit. The best of these was selected, and this was the berry now designated Iceberg and offered for $2,500. (It was bought by John Lewis Childs, and distributed throughout the United States, but it never aroused more than amateur interest.) Burbank went on to make further selections from the Iceberg, and a variety called "Snowbank" was evolved. Its fruit is said to have been "pearly white, tinged yellowish," and it was probably the closest thing to a truly white blackberry ever produced. It was neither hardy nor productive, and as one of its critics said, the public did not want white blackberries; it wanted black blackberries. However that may be, as Jones observes:

> It is an excellent example of the variation that can be induced by crossing. Thus, a yellowish white berry was made very nearly white by crossing with a fruit entirely black. Since Mendel's principles have become known, plant breeders talk freely of unit characters and modifying factors. The main difference between black and less colored blackberries, judging from similar differences in other plants, is transmitted in a relatively simple manner. Just as is that inherited something that makes one flower red [and] the other white in the same species. Burbank found that his white blackberries when inter-pollinated came true from seed in color of fruit, showing this character to be recessive. Since this experiment in crossing and from that, the building up of a trait that was not fully expressed in either parent, was carried out before anything was heard of Mendel, the white blackberry has more importance than merely a garden novelty.[4]

There was also a spectacular offer of lilies. "All the earth is not adorned with so many new ones as are growing at my establishment," Burbank avers.

> Though not having a very large stock of some, yet the number of varieties being so great, I will offer the control of some very handsome, hardy ones at from $250 to $10,000 each; and, having a multitude of other new things to absorb my attention, will offer *all these lilies* to any responsible party or parties for $250,000, including all unbloomed hybrid seedlings and all the hybrid seed produced this season; this offer holds good only to November 1, 1894.

He could hardly have been serious. No buyer was going to come forward to claim his hybrid lilies, however numerous and fine, at so astronomical a price. And putting a shutoff date on the offer smacked distinctly of the sort of publicity ploy associated with P. T. Barnum, then not long deceased.

Yet Burbank did do interesting work with lilies. His achievements in this area were called "monumental" by Carl Purdy, a contemporary authority on Pacific Coast lilies and himself a commercial lily grower. Burbank gave particular attention to the native Californian species, crossing the leopard lily, *Lilium pardalinum*, with *L. parvum*, *L. parryi*, and *L. washingtonianum*. A particularly difficult cross was made between *L. humboldtii*, a fragrant white with large spotted flowers, and *L. parryi*, which is tall and slender with clear yellow blooms. The hybrid, described as "a delightfully fragrant, pure yellow lily," propagated vegetatively, was named *Lilium burbankii* after its originator by a New York nurseryman, and was long grown in England and Holland.

Praise flowed in from horticulturists in other parts of the country. "Allow me to thank you most heartily for a copy of your unique presentation, *New Creations in Fruits and Flowers*. It is a rare, rich feast, and I congratulate you upon your marvelous accomplishments," wrote Professor T. V. Munson of Texas, himself the originator of a number of varieties of peaches and grapes. "Your pamphlet . . . is exceedingly interesting and I prize it," wrote G. B. Brackett, secretary of the American Pomological Society. "Allow me to thank you for your catalogue of *New Creations*," said Professor C. C. Georgeson of the Kansas State Agriculture College. "It is the most interesting catalogue I have ever received. Every one of your new plants is a monument in your honor."

Endorsement came from abroad, too. "We cannot say enough in praise of the Burbank Plum; it is superior to all other varieties; the most fastidious cannot find fault with it. This is without doubt the best and most profitable Plum in cultivation," wrote D. Hay & Son, of Auckland, New Zealand. "The Gladioli received by me last spring fully supported your claims for them. Thanks for your catalogue. It is a revelation to me," said H. H. Goff of Ontario, Canada, another professional nurseryman.

In 1894 the Santa Rosa *Democrat* noted:

> Fred C. Smith, Horticultural Commissioner for the South Australian Government, accompanied by Wm. Brooke, also of Australia, and

> Mr. Cillie of South Africa, spent Tuesday in this city. One of the chief objects of their visit was to meet Luther Burbank, who is one of the best-known horticulturists. Mr. Smith says his works are extensively read in Australia, and are looked upon as eminent authority. He was as much pleased with the man as with the author.

The "works" in question can, at that date, only have been Burbank's sales literature, numbering then almost forty separate items. This would tend to indicate the seriousness with which the *New Creations* catalogues in particular were received. But, on the other hand, perhaps Mr. Smith did not really know what he was talking about, or the reporter got him wrong.

The press was always only too ready to laud and exaggerate Burbank's achievements. Local papers such as the Santa Rosa *Democrat* and *Republican* waxed ecstatic. *Orchard and Farm* dubbed him "the wizard of horticulture"—a title that was ultimately to prove uncomfortable. In 1895 a Santa Rosa newspaperman named H. W. Slater reported sensationally that Burbank had contracted with a customer to produce a hardy tea rose for $5,000. This feat does not seem to have been accomplished, but Burbank had been working with roses since the mid-eighties, if not earlier, and ultimately introduced numerous varieties. Ever since George Washington crossed the wild Prairie rose with cultivated European roses at Mount Vernon, producing the variety that came to be known as Mary Washington, this flower has had special attention from American breeders. Over 600 varieties had been originated in the United States by 1920. Here, at least, there was no shortage of competition.

Typically, Burbank did not know the exact ancestry of his rose hybrids. "The parents . . . being themselves hybrids of complicated ancestry," he confesses, "it is obvious that the pedigree in a few generations became so complicated that if one were to attempt to trace them there would be little time left for any other experiments . . . so I have contented myself with watching for results among the hybrid progeny of my roses of multiple ancestry."

W. Atlee Burpee paid $500 for a rose named the "Burbank" in June 1896. A cross between the Bon Silene and a seedling of the Hermosa, this was awarded the gold medal as best budding rose at the Louisiana Purchase Exposition in St. Louis in 1904. A perpetual-blooming pink rose, it was described by its originator as hardy, vigorous, and disease-resistant. It survived as a variety for many years and was reintroduced by Stark Brothers around 1936.

Tens of thousands of Crimson Rambler seedlings were pollinated by Burbank from other varieties commonly grown in California. From this experiment in mass hybridization came the Corona, described as a "large single-flowered pink with thick waxy petals, that would keep fresh for two weeks." The other parent in this case seems to have been the Cherokee. So proud was Burbank of this variety that he grew it in the place of honor over the doorway of his house.

One of Burbank's more widely touted productions was a remarkable blue rose. "This unusual color character evidently was not stable, however," says Howard, "for no blue variety was announced."[5] (It is reported to have been a small, lavender blue cluster rose.)

Burbank's domestic circumstances now took a considerable turn for the better. In 1896 his divorce was granted, and the uncongenial Helen returned to Denver, passing out of his life forever. He had not cohabited with her for four years, he charged in suing for divorce. She had bedeviled him until he was a nervous wreck. He was a sick man and unable to carry on his business affairs. He was willing to make his wife a cash settlement of $1,800 and allow her to take all her house furniture if he could be rid of her.

A neighbor, whom the court called as a witness, substantiated much of Burbank's story, including the episode of the blackened eyes, and the fact that the plaintiff was domiciled in the barn. Helen did not appear, and her attorney did not deny the charges. She was, no doubt, as anxious to be free as Luther. They were thoroughly incompatible. Helen had made her attitude quite plain. Luther's mother, she is quoted as saying, was "a vile serpent, an old vicious cat." And she went further than that. "Luther and all his relations were a nest of cats and snakes and low-lived dogs."[6]

The settlement offer was accepted, and on that note Helen's role in the Burbank story comes to an end. He and his family were so anxious to be done with her that she is mentioned in none of the books produced by Burbank with the aid of his various ghost-writers and editors, not even in his "autobiography," *The Harvest of the Years*. Emma's little volume stops conveniently short of Helen's arrival on the scene and makes only the briefest admission that "Burbank was twice married, but left no children." (Emma did not, it seems, get on with either of the Mrs. Burbanks.) Were it not for Walter Howard's patient detective work in the two decades after Burbank's death, there would scarcely be any other record of Helen's existence. The first marriage, to the Burbanks, was an episode to be forgotten. Time passed, and eventually few people knew that there had ever been a "first Mrs. Burbank." Helen, the Denver "adventuress," was exorcised by a family conspiracy of silence.

She is heard of on only one other occasion, and it does no kindness to her memory. It seems that one morning around 1900, Burbank was opening and reading his mail when he suddenly burst into laughter. His secretary gazed at him in surprise, and he explained that the letter he had in his hand was from a Denver man who had married Helen and now wanted Burbank's advice, in his capacity as ex-husband, on how to get rid of her.

In 1897 Burbank was ready to introduce his "Improved Beach Plum," one of the things that Hugo De Vries was to find so impressive in 1904. The beach plum, *Prunus maritima*, is a wild fruit that grows in the sand dunes along the coast from Virginia to New Brunswick, doing best in Massachusetts and New Jersey. The bushes grow low to the ground in the dunes and are frequently covered by the shifting sands after they have set fruit. The branches have to be pulled from the sand to harvest the fruit, but it apparently ripens well under these conditions and is protected from the usual insect blemishes.

On Cape Cod, local families had long been used to collecting beach plums for jelly and preserves, which found a ready market with summer visitors. Burbank would have been familiar with the species as a boy, and it was exactly the sort of thing that would attract his questing mind, always on the lookout for plants that had not yet been subjected to the process of human improvement.

There is no information as to whether the Improved Beach Plum offered in 1897 was a hybrid or a selected seedling, but it is described, in a catalogue of that year, as a "compact, handsome tree, enlarged in all respects and the fruit is a beautiful purple, dotted white, with a white bloom and delicious to eat fresh from the tree, not having a trace of the original bitter taste. . . . Flesh deep yellow, freestone. . . . Ripens with the common beach plum. . . . Trees bloom . . . later than any other plum."

After his work at selection of the beach plum was complete, the result was crossed with one of his Japanese hybrids, a task that presented some difficulty, for

> it had been generally supposed that the Beach plum could not be pollinated from, primarily because it blooms very late, after most cultivated plums have passed that stage. But I was not to be thwarted by this obstacle, and used here a device I often used both before and since. What I did was to search out the latest bloomers among my cultivated Japanese hybrids and determine ahead of time that their blossoms would still be capable of fertilization when the very earliest Beach blooms were opened. By cross-pollinating between the latest of the one and the earliest of the others I effected a union that many horticulturists had given up as hopeless.[7]

The end product of this process of selection and hybridization was the "Giant Maritima," not introduced until 1905, but supposedly capable of bearing fruit as much as eight and a quarter inches in diameter.

In 1897 Burbank's gross sales exceeded $16,000. The following year, he acquired his first important foreign client, H. E. V. Pickstone of South Africa, a prominent nurseryman based at Simondium in the Cape Province. Pickstone, who would remain a lifelong customer, was particularly interested in the Japanese plums. These, notably the Santa Rosa and Beauty, came to form the basis of a considerable industry in South Africa. The Santa Rosa remains the most important export variety there, along with Gaviota, while Satsuma is grown for the processing industry, and Beauty continues to be cultivated on a smaller scale (as of 1974).

Further supplements to *New Creations* appeared in 1898 and 1899.[8] The 1898 supplement quotes a slew of testimonials on its cover, indicative both of Burbank's swelling reputation and the need to fend off his critics. The Santa Rosa *Republican* declared him "the man who has done more to give the world some idea of the possibilities of Nature's work in the world of horticulture than any other of the great scientists of the age." The *Gardener's Chronicle* of London, England, remarked, a hint chauvinistically: "The statements made to the number and variety of Mr. Burbank's productions are so astounding that some might be supposed to consider them as so many flowers of rhetoric, such as we are accustomed to from the other side of the Atlantic, were they not authenticated by competent observers of established repute." The San Francisco *Examiner*, keystone of the young William Randolph Hearst's rising empire, waxed poetic: "Not Ruskin, perhaps, nor Tolstoi, with all their love and study of human nature, has learned a deeper wisdom than has come to this patient, studious man who has given his love to the strange, silent forms of life we call vegetable, and which play their parts so quietly that to many they are insignificant and half forgotten."* To the San Francisco *Call* Burbank was, quite simply, "the Edison of horticultural mysteries."

Burbank now tendered his own name as a warranty of excellence. HOW TO JUDGE NOVELTIES—LOOK TO THEIR SOURCE, he exhorts readers of his catalogues. The Edison of horticultural mysteries, we are evidently to understand, is no common seed peddler. "What Shakespeare was to poetry and the drama Luther Burbank is to the vegetable world," an unidentified admirer is quoted as saying. This sort of puffery, appearing in Burbank's own catalogues, was to prove increasingly galling to his competitors. In the years to come, they would lose few opportunities of taking him at his word, referring his failures back to their source, no matter what other hands had intervened, and pointing up his exaggerations with malicious glee.

Meanwhile, his star rode high. In 1899 the *Call* published an interview with Eugene Woldemar Hilgard, the doyen of California agriculture, in which Burbank was sweepingly endorsed. "He seems to have an especially swift and accurate judgment respecting undeveloped seedlings," said Hilgard. "Many people have spent their lives in cross-fertilizing desirable varieties, but without accomplishing nearly as much as Mr. Burbank has done since he came to California. His delicate perception of minute plant variation amounts to positive genius."

As the century drew to a close, Burbank, at fifty, found himself firmly established in the public eye as "the wizard of horticulture."

* The city editor of the *Examiner*, Arthur McEwen, said that when he looked at the front page, he said: "Gee, whiz!" When he turned to the second page, it was: "Holy Moses!" On the third page, he roared: "God Almighty!" (John Tebbel, *The Media in America* [New York: Thomas Y. Crowell, 1975], p. 265).

10

FRUITS AND PENALTIES OF FAME

By 1900 Burbank's plums were grown all over the world. His flowers were listed in virtually every seed catalogue that appeared. The Burbank potato had become one of the leading varieties in the United States, especially on the West Coast. "Newspaper and magazine editors, special feature writers, garden club lecturers discovered this man who, to them, was doing something new," Jones observes. "Unusual interest in the breeding of new plants and animals had been awakened in scientific circles by the rediscovery of Mendel's laws of inheritance. Biologists in Europe and America saw that Burbank was doing some remarkable things of real scientific interest."[1] At the same time, fame had its disadvantages. In his early years, he had had time to test his introductions carefully before putting them on the market. Now, with increasing claims on his time that had nothing to do with his work, he gave them out faster and faster.

One of his best-known and most lasting accomplishments was the creation of a new variety of daisy. "While there may be room for skepticism regarding his claims about certain productions," says Howard, "there should be no honest doubt about the Shasta daisy. Though the details of the ancestry are incapable—at this late day—of scientific proof, the fact that he did produce the Shasta daisy by breeding, essentially as claimed, is attested by Professors E. J. Wickson of the University of California and Hugo De Vries of the University of Amsterdam. Both these writers viewed the evidence on the ground."

It has been suggested that Burbank simply improved a perennial wild ox-eye daisy, *Chrysanthemum leucanthemum*, native to the Old World but found by him in the vicinity of Mount Shasta in Northern California, or a garden form of the wild *Chrysanthemum maximum* of southeastern Europe. It seems most probable that starting with a breeding stock of wild daisies, probably from New England, he introduced pollen from two European species, carefully selected the resulting plants, and finally crossed his hybrids with the Japanese species, *Chrysanthemum nipponicum*, to achieve the pure white flower he was aiming at. This work was completed by 1901, when it was announced in a four-page circular as "the latest floral wonder." The choice of name was a happy one, and the quadruple hybrid was hardy and adaptable. "As a type of daisy," Howard says, "the Shasta promises to survive indefinitely. Improvements on the original and its siblings, although given distinctive variety names, will doubtless always be referred to as Shasta daisies."[2] Burbank later distributed new selections as Alaska, California, and Westralia Shastas. Shasta became the name of a new type of daisy rather than of a single variety. A hardy perennial, it does well in a variety of climates, is easy to grow, and makes a good cut flower. It has retained its popularity to the present day. As with many of Burbank's introductions, the new daisy was being sold (he never disposed of it outright) before it was formally announced: a note of March 1900 records a sale of Shasta daisy hybrids for several hundred dollars.

Burbank was working hard—too hard. On August 20, 1901, he informed Wickson that

> Having overstrained myself both mentally and physically for some months I am now prostrated from this overwork, and am obliged to be at the Sanatarium at Altruria, Sonoma County, most of the time perhaps for a week or two, probably not longer. I notify you so that if you desire, you can come to the Sanatarium if you wish to see me, just a little ways out of Santa Rosa, getting off at Fulton, where there is a stage. . . . I can talk with you all you want at any time, as I am not allowed to do much else.[3]

If there is a distinctly wistful note to this letter, he must have been considerably cheered when, in the September 1901 edition of

World's Work, a new monthly magazine established by Walter Hines Page and published by Doubleday, which billed itself as "a history of our time," he received what was probably his earliest solid endorsement by a prominent scientist (if we except Wickson's encomia in the *Pacific Rural Press*). The *World's Work* piece was entitled "A Maker of New Fruits and Flowers" and was by none other than the great American horticulturist Liberty Hyde Bailey (1858–1954), who had visited Burbank not long before. From such a source, it was bound to carry a good deal of weight.

Dean of the agricultural college at Cornell University and an internationally famous botanist,* Bailey was also widely known for the books on practical horticulture and agriculture he wrote and edited. His assessment, which was both fair and unexaggerated, must have come as balm to Burbank's soul. "Luther Burbank is a breeder of plants by profession," Bailey wrote, "and in this business he stands almost alone in this country." He drew attention to the harm done Burbank's reputation by calling him "the wizard of horticulture," inasmuch as

> This sobriquet has prejudiced many good people against his work. Luther Burbank is not a wizard. He is an honest, straightforward, careful, inquisitive, persistent man. He believes that causes produce results. His new plants are the results of downright, earnest, long-continued effort. He earns them. He has no other magic than that of patient inquiry, abiding enthusiasm, an unprejudiced mind, and a remarkably acute judgment of the merits and capabilities of plants.

There follows a simple, straightforward account of Burbank's working procedures:

> From an entire tree he will pick such proportion of flowers as would be likely to fall from natural causes. The remainder, numbering hundreds, he will cross. Before the flower opens he cuts off the petals. Thus the bees are not attracted, and they have no foothold. Then he applies the pollen with a free hand. This pollen is usually collected the day before from flowers that are picked and dried. All the seeds resulting from the cross are sown. Of a thousand seedlings, a dozen may be promising. These are saved, and perhaps they are crossed with some

* According to one account, it was Bailey who, in copying a reference to Mendel's paper from a German bibliography of the literature on plant hybridization, originally led De Vries to the rediscovery. Bailey had not actually seen Mendel's report himself.

> other plant. Again the seeds are sown; and thus the process continues until a desirable form is secured, or until it seems to be futile to carry the experiment farther. The judgment as to what will likely be good and what bad is the very core of plant-breeding. In this judgment Burbank excels. Not to many men is given this gift of prophecy. Burbank calls it intuition. He cannot explain it any more than another man can explain why he is a good judge of character in human beings. Long experience and close observation have directed and crystalized this faculty of his, until it is probably as unerring as such faculties can be.
>
> Burbank loves all plants. He has worked with fruits, vegetables, flowers, grains. A strange plant in the fields at once attracts his attention and he tries to cultivate it, even though he may not know its name. His flowers and other quick-maturing things are usually grown in long, scrupulously tilled rows. Fruit trees have so long a period from seed to fruit that cions are taken from them when one or two years old, and these are grafted into the tops of bearing trees. Thereby he secures fruit sooner. In one tree there may be scores of kinds of fruit in bearing. Of most fruits he expects the graft to bear in two or three years from the seed. At the same time he may allow the original seedling to remain, thus securing two sets of the same plant with which to work. The fruit trees are planted very close in rows, and as soon as any plant proves to be worthless it is removed, and another may be planted or grafted in its place. The rows soon come to be collections of the most unrelated curiosities.

Bailey had clearly fallen under the spell of Burbank's personal charm. "You feel his kindly and gentle spirit," he says, "and before you know it you love him." He is remarkably uncritical of Burbank's failure to keep proper records and to maintain precautions against self-pollination. He seems almost to exonerate it entirely:

> Mr. Burbank no longer makes any serious effort to keep a written record of his crosses. He remembers the parentage. In many cases he applies the pollen of two or more kinds of plants to one flower. He does not know which pollen will "take." Neither does he always remove the stamens from the crossed flowers, as we are always advised to do in order that the plant may not be self-pollinated. In practice he finds that this precaution is usually unnecessary, for the pistil is likely to refuse pollen from the same flower. When the seedlings come up, he can tell what the cross was; or if he cannot, it matters little, for he is not making his experiments primarily for the purpose of accumulat-

> ing scientific records but in order to obtain definite results in new varieties. Yet, so careful and acute are his judgments that one places great confidence in his conclusions as to parentage; and many times he makes crosses with every scientific precaution. I must confess I was skeptical as to the existence of the "plum-cot," or the cross between the plum and apricot; but now that I have seen many of the trees in bearing I am fully convinced that he has produced plum-apricot hybrids. The marks of plums and apricots are too apparent in the fruits and trees to be doubted.
>
> Mr. Burbank gets unusual hybrids because he crosses great numbers of flowers and uses much pollen. He is skillful in the technique. He also dares. He has no traditional limitations. He knows no cross that he may not attempt. He has not studied the books. He has not been taught. Therefore he is free. The professor of horticulture would consider it beyond all bounds of academic and botanical propriety to try to cross an apple on a blackberry; but Luther Burbank would make the attempt as naturally as he would dig a new lily from the fields.

The visitor also cast an interested eye over Burbank's bookshelf. He found it sparsely stocked. Aside from Darwin, he notes, Burbank's chief guides in technical botany were Asa Gray's *First Lessons in Botany and Vegetable Physiology*, a high school textbook published in 1857, and *Field, Forest and Garden Botany* (1868). He observes that

> some philanthropist could render a good service to mankind if he would endow this experimental garden and allow its proprietor to devote his whole energy to research. The best fruit-growers of California prize Burbank's work and are confident that his varieties will win. In visiting his place, one feels regret that scientific record is not being made of these rich experimental results. Mr. Burbank shares in this feeling, and he would welcome any careful and sympathetic student who should essay to make a permanent record of the work as a contribution to scientific knowledge. His place is an experiment station of the best type. His work makes for progress.

There was no great theoretical gulf between the two men. A botanist of the old school, Bailey saw Burbank's practical goals and seems on the whole to have favored them against "academic and botanical propriety." He was a bit of a maverick himself. At that time, moreover, his views on the mechanism of speciation were not far removed from Burbank's. "Professor Bailey's philosophy," Edwin Conklin observed in 1896,

> was neither strictly Lamarckian nor Darwinian, although in general it leaned to the former; it was rather *sui generis* and might be called Baileyan. He maintained that variability is the original law of organisms, that like no more produces like than unlike, but that mutability is a fundamental and normal law, while heredity or permanency is an acquired character. The organism is shaped by its environment, and nature eliminates the non-variable and favors the survival of the unlike.[4]

The difference between the two men was that whereas Bailey was prepared to amend his views in accordance with the new knowledge that flowed from the rediscovery of Mendel, Burbank remained stubbornly true to his original opinions.

Edward Wickson, head of the Department of Horticulture at the University of California and subsequently for many years dean of the College of Agriculture and director of the Agricultural Experiment Station at Berkeley, was also the editor of the *Pacific Rural Press*, probably the most influential of all the horticultural journals on the West Coast. Simultaneously educator and journalist, Wickson, a portly man with a handlebar moustache, was for decades one of California agriculture's most effective publicists. In 1901 he was collecting material for a series of articles that would serve to make Burbank more famous than ever.

Among those Wickson approached was Judge Samuel F. Lieb, a prominent San Jose attorney (he was a member of the original board of trustees of Stanford University and an intimate friend of the Leland Stanfords), who had been a Burbankian for five or six years at the time—after having initially been inspired to visit the wizard by an article in the *Pacific Rural Press*. From this visit, Lieb informed Wickson in a letter dated October 11, 1901, "a friendship sprang up between us which makes us like brothers." He knew Burbank as well, he believed, as any man alive, and

> I have never known a nature more full of absolute sweetness. He is absolutely honorable in every way and is honest to a fault. He lives what is termed in the parlance of the day a strenuous life, far too much so for his physical endurance. He is an intense man, a man who

> carefully plans for results and then works for their fulfillment with a patience that exceeds that of Job himself. It may be a question of years to arrive at a single result. Necessarily before arriving at success in seeking to accomplish a given result, he must meet with many failures, but nothing seems to daunt him until success finally crowns his efforts.

Lieb goes on to describe some of the things Burbank was attempting at the time. Although the experiments he speaks of never amounted to anything, they give a further insight into the immense range of Burbank's endeavors, extending over the whole spectrum of cultivated plants, far beyond the scope of an ordinary nurseryman.

> Last summer I noticed a patch of parsnips at his experiment grounds. I was astonished that he should be devoting his attention to a thing of that kind, and asked him what it meant. He told me that he was trying to evolve a new plant for sugar. The sugar beets have developed so many diseases and so many objections have been found to them that he is trying to get something better. These diseases and objections are all obviated in the parsnip, if it can be gotten full enough of sugar to answer the purpose. He has already improved the variety in that direction so much that it is just on the verge of being successful. I have not the slightest doubt that he soon will have it so that by analysis it will be found to contain as much sugar as the beet, and doubtless then it will be exclusively used instead of the sugar beet.

This attempt at developing a sugar parsnip came to naught,* but it bespoke a far-reaching imagination. What is also apparent from Lieb's account is the extent to which the public already looked on Burbank as a miracle worker who could turn out new varieties to order. Lieb speaks of requests for a new cotton, a new sugarcane, a new, improved coffee. Burbank could not undertake these projects, but Lieb is in no doubt that he could do it if he had the time. His enthusiasm for his hero knows no limits.

> What he will succeed in doing before he dies the Lord only knows. I do not. I only hope the good Lord will preserve him to us for the good of mankind as long as is possible, and "May God bless him" is my constant prayer. . . . I can but say that a genius like this is only found

* Burbank announced only one variety of parsnip—a selected strain of Hollow Crown called Imperial Hollow Crown—and that was in 1919.

> once in a century, and the general community, who alone is to reap the benefit should see that nothing of that genius is lost. In my judgement the United States should take such a man and put him to work with every appliance and facility which can be afforded him, to make new productions and better old ones, all for the common good of its citizens. It should say to him, "We have employed you for your genius. Here is all the land you need for your experiments, all the water you need for its irrigation, all the laborers you need to carry on the work, all the foremen you need to carry out the details. What we want you to do is to give us the benefit of your genius to the fullest possible extent.[5]

This uncompromising faith is all the more extraordinary in that Lieb was himself a fruit grower and far from technically ignorant of horticulture. Burbank, Howard remarks, had given him a "sign" by successfully prejudging two sets of seedling fruit trees, and he was henceforth a true believer. "Except ye see signs and wonders, ye will not believe," says St. John. Evidently, Burbank was quite capable of providing them.

The first of Wickson's four articles on "Luther Burbank, The Man, His Methods and Achievement" appeared in *Sunset* magazine in December 1901. The following three appeared in the February, April, and June 1902 issues. They were all subsequently issued in book form by the Southern Pacific Company, which was then publisher of the magazine, with a foreword by *Sunset* editor Charles Sedgwick Aiken comparing Burbank to the "soldier hero, who at the outbreak of the Spanish war, carried this Nation's message to Garcia." Wickson himself waxed intemperate in the extreme, in language that is the more astonishing coming from a figure of his academic respectability:

> For such a gifted seer neither weird altar fires, nor incense cloud, nor ecstatic state could add to insight. He could hear the "still small voice" without preparatory earthquake or whirlwind. Like David of old he could do his work with smooth pebbles from the brook; and he cast aside the elaborate armament of his scientific brethren lest it should impede his movements.[6]

There is more than one crack at the scientific establishment. "He has worked through a country not yet officially surveyed, above the pathway of the contemporaneous scientists," Wickson asserted, "and it is not wonderful, then, that they should fail to recognize him for a time." Critics implied that Burbank was making a travesty of science. He had patiently borne "the burdens of distrust and misapprehension which fall usually to the lot of those who extend the frontiers of human knowledge." Future generations would, however, recognize him as "a lone star glowing in the horticultural horizon."

In the long run, this sort of thing could do Burbank no good, but in the short term it was amazingly influential. Botanists who had the fortitude to say that many of the things Burbank claimed were impossible, Jones observes,

> were immediately discredited as being professionally jealous. No one has ever said that American biologists were jealous of Mendel, a Catholic monk whose name is heard more often in scientific circles than Burbank's; they were not considered to be jealous of Darwin who, like Burbank, was not a professional biologist; they were not jealous of Galton who founded the eugenics movement; but whenever any professional botanist or agricultural investigator stated anything in disparagement of Burbank he was immediately put down as envious of his reputation.[7]

True, the botanists were not necessarily right. The existence of amphidiploids, which could conceivably account for some of Burbank's more extraordinary hybrids—as Jones himself was the first to recognize—had not yet been established or even suggested. But in the face of the threat to their professional integrity that Burbankism presented, the scientists were bound to hang together. Almost inevitably, the baby got thrown out with the bathwater.

Initially, however, reactions were surprisingly good. "Dear Prof. Wickson," wrote G. B. Brackett, pomologist of the USDA Bureau of Plant Industry, on December 27, 1901:

> I am just in receipt of a copy of the magazine "Sunset," containing an article on "Luther Burbank, The Man, His Methods and His Achievements," for which I sincerely thank you. It delighted my soul to see that photo of yourself and Mr. Burbank in characteristic conversation. . . . No pen can over rate the great work of Mr. Burbank, and I am glad you have given to the world so just an estimate of the worth of the man we both so greatly admire. He is easily the foremost horticultural experimenter. I prize this edition of Sunset greatly. A Burbank edition for the Christmastide was a happy idea.

In May 1902, Arthur A. Taylor, proprietor of the Santa Cruz *Surf*, wrote: "Mr. Burbank is a miracle worker and is as yet as much without honor in his own country as was the Man of Galilee." Here were obviously two more disciples. (Taylor's letter suggests, moreover, that Burbank was far from alone in the "Messiahship idea.")

A note of dissent was sounded by one Thomas Lyon: "I have just read with interest but with great disgust your second paper on Mr. Burbank's work with plants," he began.

> You spoiled your work by constantly using the "jargon of the schools" in place of words that could be understood by your least educated reader. . . . As you well know Burbank has evolved no new method. He is merely working well and honestly in a very old and well trodden road and hence there is no use in trying to cover his work with a halo that does not belong to it. He has done good work and has achieved good results and deserves credit therefore but he is but one in the great list and deserves no more than his work is really worth. I am sure if he comprehends the meaning of your words his modesty will be much mortified. . . . Burbank was not working for fame but for dollars.

Wickson forwarded this letter to the sage at Santa Rosa, and far from being modestly mortified, the latter responded with a sneer at the "crisis of wind-colic which affects our 'Grape-nut-brained' friend and adviser who calls himself a Lyon." Burbank went on:

> Judging from the muddy tracks of this specimen, I should confidently place the thing under the head of Cimex Lectularius, or Bedbug in language less polite, rather than Lyon as alleged. This specimen, however, appears to be a degenerate microcephalus form of the parasite. Did the thing get the idea you were writing for a Sunday School Journal, or a Kindergarten Paper? Well, these things do irritate in spots, but are not alarming. These little fellows don't come out in the daylight much.[8]

He was quite prepared on occasion to carry the fight into the camp of his critics. Speaking at the Floral Congress in San Francisco in 1901, he declared contemptuously:

> The chief work of the botanist of yesterday was the study and classification of dried, shriveled plant mummies, whose souls had fled, rather than the living, plastic forms. They thought their classified spe-

cies were more fixed and unchangeable than anything in heaven or earth that we can now imagine. We have learned that they are as plastic in our hands as clay in the hands of the potter or colors on the artist's canvas, and can readily be molded into more beautiful forms and colors than any painter or sculptor can ever hope to bring forth.[9]

There was sense in what Burbank said, perhaps, but such vainglorious phrasing could scarcely aid his cause.

Burbank's friends, Wickson and Lieb in particular, were constantly urging him to perpetuate his skills and knowledge, either by writing or by teaching. But he had little time or inclination for either. Nonetheless, in 1902 he did send a short paper entitled "The Fundamental Principles of Plant Breeding" to be read at the first International Conference on Plant Breeding and Hybridization held that year in New York, and this was later published in pamphlet form. That same year there was an attempt on Wickson's part to recruit him for the University of California. It was politely but firmly declined. "In the first place I am perfectly sure that I can do much better work for the *world* right in the harness where I now am," Burbank wrote him.

As to remuneration, that would make no difference with me, for I am making absolutely nothing now, just keeping even, anyway. It is all a work of love, or duty to humanity. But if I should look to the financial part, I have received propositions from another University* with ten times the remuneration proposed for the first year, and a promise of much more afterwards, and numerous other very valuable considerations outside of salary; but these facts make no possible difference with my decision either way.

I appreciate your very great kindness in this matter, and years ago when I was younger and not so thoroughly engaged in matters which cannot well be given up I might have thought it best to have accepted such an offer, but at this critical time when results of decades of work are ripening, and that too with comparatively little present effort, I am thoroughly satisfied that I am right in this matter, and yet I am sorry if my decision shall be antagonistic in any way to the wishes of my kind friend Wickson.

* Undoubtedly Stanford, where in the persons of President David Starr Jordan and Professor Vernon Kellogg he had two ardent supporters.

That the proposal was official, and not simply Wickson's own idea, is confirmed by a note to the latter from University of California President Benjamin Ide Wheeler, dated September 24, 1902. "I share your regret [at Burbank's refusal]," Wheeler remarks.

Burbank was more fortunate in his friends than perhaps he realized. In 1902 the Carnegie Institution of Washington was set up by millionaire Andrew Carnegie (1835–1919) with a multimillion-dollar endowment and a charter "to encourage in the broadest and most liberal manner investigation, research, and discovery, and the application of knowledge to the improvement of mankind." There appears to have been something of a concerted effort, perhaps guided by Wickson, to urge support for Burbank, whose name appears in the files of the institution as early as March 1902. In November 1903 a formal application for a grant in aid of his work was submitted by David G. Fairchild for the Evolution Committee of the Botanical Society of Washington. In a letter dated December 9, 1903, President Wheeler notified Wickson of its failure, advising him of a "communication from the Carnegie Institution which I know you will read with regret, as I did." Citing the "many applications received to date for aid in support of or furtherance of various scientific projects," Charles D. Wolcott, secretary of the institution's executive committee, wrote: "It has been found impossible to comply with these requests in most cases. I regret to advise you that in the case of L. Burbank it has not been found possible to make the grant requested."

L. Burbank would get his grant eventually, but he would have to wait until bigger guns could be dragged up on his behalf.

Despite his pious protests about "just keeping even" and the failure of the first grant application, he was making money. These were years of steady sales. In 1902 John Lewis Childs paid $1,100 for a "phenomenal hybrid berry" and plants and seeds of Crimson Winter rhubarb. In November 1903 the Oregon Nursery Company bought the "Miracle" plum for $2,500. This was the so-called stoneless plum. The buyers advertised widely, and it was extensively planted at the time.

There were, of course, numerous smaller sales, the combined value of which at least equaled the big ones. Toward the end of

1903, or early in 1904, Burbank was sufficiently well off to buy his first car: an Oldsmobile. The tokens of recognition were also piling up. At the opening meeting of the American Breeders' Association at St. Louis in 1903, he was unanimously elected to honorary membership. In the early proceedings of the association, a number of articles by him appear, under the titles "Heredity," "Right Attitude Toward Life," 'Another Mode of Species Forming," and "Evolution and Variation with the Fundamental Significance of Sex."

The Carnegie grant, though denied, was obviously still in the offing, for it is mentioned without comment in a note of January 1904, apparently typed by Burbank himself. Wickson was probably busy pulling strings and may have given him some encouragement around this time. The president and chairman of the executive committee of the trustees of the Carnegie Institution was Daniel Coit Gilman, Wheeler's predecessor as president of the University of California. An influential member and later chairman of the institution's finance committee was D. O. Mills, a prominent San Francisco banker and merchant. Another member of the finance committee, Lyman J. Gage, formerly secretary of the treasury, could be regarded as distinctly friendly, as could Judge William W. Morrow of the board of trustees, who had lived in Santa Rosa and married there. It was a common rumor at the time, Howard says, that California would have its share of the Carnegie money and that the "California members" among the trustees were bringing pressure to bear on the others to this end. With the support of Wheeler and other Burbank enthusiasts such as David Starr Jordan, president of Stanford University, Wickson must certainly have believed that he would ultimately be able to swing the board in Burbank's favor.*

In July 1904, with the appearance of an article entitled "A Maker of New Plants and Fruits" in *Scribner's* magazine, a new champion joined the lists. W. S. Harwood was a professional journalist and popular science writer. Prolific and given to extravagant hyperbole, he would do more than anyone else to spread the Burbank myth and at the same time to ruin Burbank's reputation

* Burbank had many proposers, among them Louis Blanckenhorn, a Los Angeles stockbroker, who put forward his name in a letter dated May 14, 1902.

among serious horticulturists and workers in the new science of genetics.

Of course, there were always those who did not fall under Burbank's spell. In 1904 he had a visit from his South African customer H. E. V. Pickstone, who spent the day with him. "I was disappointed with his personality," Pickstone later noted. "I found him too much of an egoist. . . . I do not think he can be considered a great man from any angle, but he was not a fakir; I believe him to have been quite honest but perhaps unbalanced." For all that, he remained a customer to the end of Burbank's life, though concluding in somewhat jaundiced fashion as of 1938 that "the only variety of his that has proved of commercial value is the Santa Rosa plum."[10] This was hardly true even in South Africa, so it may perhaps be supposed that Pickstone was venting some undisclosed grievance. These turn-of-the-century nurserymen took offense easily.

In January 1905, Wickson was again on the lookout for biographical material—now apparently with the aim of telling the story of Burbank's life in "simple, bold and stirring English." Having been just about to leave on business when Wickson's letter broaching the subject arrived, Burbank delegated his reply to his faithful secretary, May Benedict Maye, who had worked for him for the past five years or so. Miss Maye was evidently a woman of character and sense and a fervent Burbankian besides.[11] Noting that "reporters have camped on his trail for some time lately. . . . They have made this their Mecca from before breakfast until as late as midnight on some occasions, and life has been one round of filling their demands for 'Copy,' 'Stuff,' Material and Photographs," she listed the six most important things Burbank had done "from my standpoint and to the best of my ability with some suggestions that I have had from the Boss himself." These were:

1. *Popularizing horticulture* and placing NEW IDEALS of horticulture before the people, giving it a CLASSICAL position before the world. (Perhaps he, more than any other one man has done away with the title "Hayseed" in connection with tilling the soil. People now realize that the scientific farmer or plant creator has a work as great as any other of the artists and can change the desti-

nies of the human race, and create *Masterpieces* that go on *re-creating themselves*. . . .)

2. The bearings of Mr. Burbank's discoveries in plant life in their relation to the improvement of the human race, and the future biological application that will be made of them in the uplifting of humanity when the laws governing heredity in plant life shall be used in human improvement.
3. Bringing scientific and practical horticulture on one common ground. In fact proving by his discoveries that the best scientific work in the truest sense of the word is also *Practical*.
4. Discovery of *New Tendencies of heredity* and the laws *governing plant improvement*, so that others may avoid mistakes and continue the work with the advantage of his experience.
5. Destroying the preconceived notions of many scientists of the past and *building new ones in their place* so that we may turn from the *old ideas* of the past to the *facts* of today. Perseverance giving place to *pliability*, etc. etc.
6. Last, the new creations in fruits and flowers, grains, nuts, etc., that are a heritage, not only to this, but to all future generations as long as man shall cultivate the soil . . . his true greatness is not measured by the actual fruits and flowers Mr. Burbank has produced. Others have made new roses, new carnations, new watermelons, new fruits—here and there all over the world plant breeders have been at work; our gardens and orchards testify to the facts. But Mr. Burbank has done more than merely by infinite patience produced these new fruits, etc. He has *unearthed* the *laws* that have governed their production and made bare truths that will *live and benefit humanity* even *if they should ever cease to eat plums or care for potatoes*.

"P.S.," she adds, "If this isn't a 'full dose' come up and we can give you some more, Boss will be home after Monday."

This exposition is especially interesting in that it lays stress on Burbank's belief in the implications of his work in human eugenics and on the idea that he had unearthed the laws of plant improvement. What these were it is hard to tell since for all the torrents of words attributed to him, he never succeeded in clearly formulating them. It is true that there were implications in his results that, had they been trusted or heeded by contemporary scientists, might just conceivably have led to an earlier understanding of polyploidy.

But even this would have depended on comparable progress in cytology.

New laws aside, however, Maye is not far wrong in according first place to his popularization of horticulture and last to his new creations. This in the end must be the verdict of history also. It is to the credit of Burbank's sharp wits (we may be reasonably sure that his secretary was only quoting him) that he perceived this, when his contemporaries—including Wickson and De Vries—failed to do so. Long after his death, agricultural scientists and horticulturists would reckon this his chief merit. As Professor Wendell Paddock, an Ohio horticulturist, wrote:

> Of course his work has been of great value: it interested people in general in plant life: it stimulated college and experiment workers to unusual activity. They could not afford to have this unlettered man forge so far ahead of them. . . . there was practically no plant breeding going on at the Stations. . . . Burbank . . . for a time was practically alone in the field.

Dr. O. M. Morris, of Washington, who had the impression that Burbank was "an egotistical maniac," nonetheless conceded that he had "been a great spur or prod to a lot of technical workers who are in federal and state institutions." Professor E. H. Hoppert of Nebraska felt "Burbank was highly overrated but . . . the publicity that he received was of some value in arousing interest in the possibilities of developing new varieties of fruits, perennials, etc." Joseph A. Chucka, a Maine agronomist, said: "It is my impression that Burbank contributed little if anything to the science of plant breeding. In my judgment he contributed a great deal to the world in arousing interest in the possibilities of plant breeding and in actually giving the world a rather large number of useful and ornamental plants."[12]

The year 1905 saw the introduction of the Giant Maritima plum, a new version of the beach plum of a few years before. In May of that year, J. L. Childs bought the Improved Giant Crimson rhubarb, a selection from Crimson Winter that was nicknamed the "mortgage lifter" by contented growers, for $1,500. In July, Burbank concluded an agreement with George C. Roeding, president of the Fancher Creek Nurseries in Fresno, California, "for sale of certain new varieties." The same month there was a major deal with John M. Rutland, an Australian nurseryman, who for $1,600 bought the Santa Rosa plum for distribution in Australia and rights in the Southern Hemisphere, including South America, to the Rutland (previously "Blood") plumcot. The Rutland had been one of a display of plum-apricot hybrids exhibited at the Pan-American Exposition at Buffalo, New York, in 1901, for which Burbank had received a special gold medal. Burbank believed it to be a cross between an apricot and the Satsuma plum, but, says Howard, it is "now definitely regarded as a large hybrid plum, with perfect flowers, but with flesh of poor quality." De Vries, who saw and tasted the results, was convinced, however, that Burbank had successfully made this cross, and other horticulturists subsequently claimed to have duplicated it.

More important than these sales, however, was the financial support of the Carnegie Institution, which was at last confirmed. In a letter dated December 29, 1904, President R. S. Woodward, who had succeeded Gilman, informed Burbank that a grant of $10,000, "to be available during the ensuing year," had been allotted "for the purpose of furthering your experimental investigations in the evolution of plants." An equal grant, Woodward noted, was contemplated "annually for a series of years, *or for so long a time as may be mutually agreeable*" (italics in original). A further grant of $2,000 was made to Dr. O. F. Cook, of Washington, D.C., "for expenses and salary while engaged in association with Mr. Burbank in investigating results, conduct of experiments, and preparation of a report," the minutes of the institution's board of trustees reveal.

The initial payment of $10,000—grant number 221—was made to Burbank in January 1905, and in July Woodward visited Santa Rosa to assess the situation. He concluded that "the best way to carry on the work would be to have it supervised by a committee consisting of the heads of our Departments of Biological Research along with the President." Moreover, it appeared "on consultation with a number of leading biologists . . . that Dr. Cook lacked their confidence to a degree which promised to make a report from him deficient in the weight essential to such a document appearing under the auspices of the Institution." Cook accordingly withdrew in October.

The Carnegie trustees treated their new protégé with something approaching trepidation. "Mr. Burbank . . . is a peculiar man," Woodward told them. "He is a most highly sensitive man, and one with whom very few men who have passed forty years of age can work without difficulty." Nonetheless, he thought it "no exaggeration to say that he has in his orchard some fruits now whose value when put on the market, to California and to the United States, will far exceed the entire endowment of this Institution." Woodward endorsed Burbank as "a man who unconsciously works by the scientific method to most extraordinary advantage." It was hoped "to train up two or three young men who can learn Burbank's tricks, so to speak."

Perhaps the trustees were already conscious of some murmuring against the grant in scientific circles, but Andrew Carnegie himself came to Burbank's support. The great philanthropist, a former cotton mill worker who had become one of the richest men in the world, obviously regarded Burbank as a fellow graduate of the school of hard knocks who deserved backing against academic pretensions.

> I think nothing you have done has attracted such general attention as your aid to Mr. Burbank. That of itself is a gain to the Institution. Now, it strikes me . . . that you have met a genius, you have met a man who has done something. I think that a man of science, who perhaps has done nothing, should go cap in hand to a genius; and you should not endeavor to harness up that genius to drive your dog-cart, because he will not drive,

Carnegie told the trustees.[13] A few days later, George Harrison Shull, a young botanist at the Carnegie Institution's Cold Spring Harbor Station for Experimental Evolution, was selected to replace Cook. As one who had not yet done anything, sent "cap in hand to a genius," his path was not to be an easy one.

The year 1905 was one of widespread publicity for Burbank, some of it from highly accredited sources. In January an article by David Starr Jordan entitled "Some Experiments of Luther Burbank" appeared in *Popular Science Monthly*, and in August the same magazine published "A Visit to Luther Burbank" by Hugo de Vries. The indefatigable Harwood published two articles, "A Wonder Worker of Science" and "Burbank's Creed," in *Century* magazine, and a third, "Luther Burbank's Achievements," in *Country Calendar*. Of these "Burbank's Creed" is of particular interest, inasmuch as it purports to give an account of his religious beliefs. "My theory of the laws and underlying principles of plant creation is, in many respects, diametrically opposed to the theories of the materialists," Harwood has him say.

> I am a sincere believer in a higher power than that of man. All my investigations have led me away from the idea of a dead, material universe, tossed about by various forces, to that of a universe which is all-force, life, soul, thought, or whatever name we may choose to call it. Every atom, molecule, plant, animal, or planet is only an aggregation of organized unit forces held in place by stronger forces, thus holding them for a time latent, though teeming with inconceivable power. All life on our planet is, so to speak, just on the fringe of this infinite ocean of force. The universe is not half dead, but all alive.

Finally, in September 1905, Harwood published the first full-blown book about Burbank. This work, *New Creations in Plant Life: An Authoritative Account of the Life and Work of Luther Burbank*, was full of typical hyperbole, so much so that Howard afterward thought it should have been entitled "New Creations in Plant Life—A Fairy Story," while prior to its acceptance by Macmillan, David Fairchild, who was apparently called in as a consultant, managed to persuade the Century Company that they ought not to publish it. "In it were claims that Burbank should have forbidden publication of or repudiated," said Fairchild, who was well disposed toward Burbank and in his own autobiography described him as "an altogether lovable person, whose intuitive sense with regard to plants was most extraordinary." He did feel, however, that Burbank allowed the people around him grossly to exaggerate the claims of his new creations, which frequently turned out less well when grown away from the climatic conditions they had been produced in. (He had, moreover, been "surprised and nonplussed" to discover that Burbank believed in clairvoyance.) Fairchild makes a particularly telling observation:

> One might describe Burbank as like Tolstoi, in that, when one was with him, one felt the strange force of his simplicity and his profound confidence in his own abilities. But, on leaving him, the impression faded, and one began to wonder wherein lay his power, for his results did not quite seem to justify his claims.[14]

No one else expresses it quite so succinctly, but many another observer seems to have been similarly affected.

11

ENTER GEORGE SHULL

The 1880s and 1890s had been a period of intense interest in the agricultural potential and development of the state of California. These years saw the beginnings of modern large-scale production, preservation, and cooperative handling of fruit. Canners were beginning to learn what varieties would process best, and the pioneer apricot growers of Riverside had demonstrated the advantages of sulfuring before sun-drying. Some idea was being gained of the preferences of eastern markets in dried and canned fruit.

In 1892 California passed a law against the introduction of fruit trees from other states, and the effect of the quarantine was inevitably to promote local nursery business. The threats of eastern nurserymen to boycott California fruit in retaliation came to naught, and the courts maintained the quarantine law, with provisions being made in 1899 for greater powers for inspectors and quarantine officers. All this had the effect of creating an ever-greater demand for California trees, and nurseries multiplied rapidly.[1]

Burbank was far from alone in the field of plant breeding, and neither did he have any monopoly on go-ahead ideas. A. T. Hatch, who had come to California as a miner in 1857, began to cultivate almonds at Suisun in 1872 and around the end of the decade originated three important varieties—IXL, Ne Plus Ultra, and Nonpareil—by selection. Francisco Franceschi, an Italian immigrant who settled in Santa Barbara and established a nursery around 1893, introduced avocado seedlings and did pioneer work with a variety of subtropical fruits, including the banana, the guava, and the mango. Franceschi even tried to introduce coffee plants, which were listed in his nursery catalogue in 1897. Others, like Anthony Chabot of Oakland, attempted to raise tea on a commercial basis. A. B. Chapman and George H. Smith brought the Valencia orange to California in 1876, and Washington Navel oranges, originating as a bud sport near Bahia, Brazil, were introduced through the efforts of William O. Saunders of the Department of Agriculture.[2]

George Christian Roeding, of the Fancher Creek Nursery at Fresno, who in 1906 assumed the marketing of such Burbank productions as the Santa Rosa plum, the Rutland plumcot, and the Royal walnut, made a personal crusade out of promoting Smyrna fig culture. Smyrna figs and the wild caprifig necessary for their pollination were imported from Turkey by Roeding's father, a German forty-niner who had stayed on to make his fortune in land development. Artificial caprification—pollination of female Smyrna fig flowers with pollen from caprifig blossoms—was initially accomplished with the aid of a goose quill, but the younger Roeding realized that if Smyrnas were to be grown on a commercial basis, it would be necessary to introduce the fig wasp, *Blastophaga grossorum*, which performed the act of fructification in the fig's native habitat. As a result of Roeding's efforts, and with the assistance of the Department of Agriculture, this was accomplished, and by 1899 the *Blastophaga* had been imported and was established in California. This was a notable achievement for the new art of economic entomology, opening the way for large-scale Smyrna fig culture. By the 1930s, California would be producing more than 20,000 tons of Smyrna figs annually.[3]

Other new lines of development found ardent enthusiasts and speculators. In the early nineties, a "great olive rush" developed. The notion spread that olives could be grown on land too poor and dry for other kinds of fruit. The Mission olive had been introduced by Junípero Serra and José Gálvez, who planted seeds brought from Mexico at Mission San Diego in 1769. In 1875 another variety, the Manzanillo, was brought from Spain, and ten years later the Sevillano and Ascolano, both mainly pickling varieties, were introduced. The number of olive trees in the state increased from less than 15,000 in 1875 to some 2.5 million in 1897. Established nurserymen rushed to supply the demand for olive trees, and no less than six new nurseries devoted entirely to their reproduction sprang up, one of the smaller of which offered half a million trees for sale in 1892.

Burbank did not neglect the olive. "The nine hundred thousand olive trees which I offer for sale this season are grown by a new process," he announces in a postcard sent out in 1888. This seems to have been purely a business proposition, and he does not appear to have given much thought to improving the tree. Growers found, however, that they could not compete economically with cheap European olive oil and other vegetable oils, and difficulties arose in pickling and shipping. Olives consigned to eastern distributors in casks spoiled en route, and early canning efforts frequently resulted in botulism poisoning. By 1895 planting had virtually come to a halt, and the hopeful olive nurseries of a few years before were in the process of closing down. (Production picked up after 1900 with the introduction of modern canning methods, and today California is the only important olive-producing state in the United States, with an annual crop in excess of 50,000 tons.)

The eucalyptus craze of 1906 was a similar phenomenon. Plantings of eucalyptus, promoters claimed, would yield as much as $2,000 per acre every six years and needed neither good land nor water. Interest was whipped up to a frenzy. Land sharks set to work subdividing desert and dunes, and the demand for eucalyptus trees reached incredible proportions, despite the warnings of university and government experts. Eucalyptus peddlers and backyard nurserymen swarmed to take advantage of the boom. Frost and lack of water killed off many of the young trees, however, and eucalyptus fever went the way of so many other California sensations.

Against this backdrop, Burbank saw fit to touch off a craze of his own. It had long occurred to him, he said, that every plant growing in the desert was either bitter, poisonous, or spiny. He had been struck by the hardiness and adaptability of cacti as much as by their potential use as cattle fodder. In the seventeenth century, Spanish herdsmen had learned to utilize cactus for feed by burning off the spines over a brush fire. The spines, which are dry and waxy, quickly burn off, while the fleshy, water-holding pads remain undamaged. Alternatively, the chopped cactus could be allowed to stand in a mass, which softened up the spines. Later gasoline torches were used to burn the thorns off.

But singeing did not always remove all of the larger spines and left the rudimentary spines imbedded in the slabs untouched. Cattle that had been fed cactus were often observed with blood dripping from their mouths, and their throats and tongues would tend to become inflamed. This was, therefore, a source of food and water cattlemen fell back on only in time of need.

But what if a spineless forage cactus could be developed?

The idea appeared to open up a whole vista of possibilities for reclaiming the barren regions of the earth. In arid parts of the United States, a vast land area, much of it given over to coyotes and jackrabbits, cacti of many species flourish. These are mainly spiny varieties, but Burbank knew perfectly well that there were varieties of cactus that were comparatively spineless. In fact, he remarks in one of his catalogues: "One of the first pets which I had in earliest childhood was a thornless cactus, one of the beautiful Epiphyllums. The Phyllocactus and many of the Cereus family are also thornless, not a trace to be found on any part of the plants or fruit."

He started to build up a collection of cacti, concentrating on the genus *Opuntia*, to which the prickly pear (*Opuntia tuna*) and Indian fig (*O. ficus-indica*) belong. Specimens were obtained from all over Mexico and from Central and South America, as well as from North and South Africa, Australia, Japan, and the Hawaiian and South Sea Islands. (As has been previously noted, all cacti, with one possible exception, originate in the New World, so that Burbank was, in effect, reimporting them in the hope that they might have sustained significant variation in their new habitats or undergone mutation.) Eight partially thornless kinds, from Sicily, Italy, France, and North Africa, were obtained for him by the plant explorer David Fairchild, while the USDA supplied him with specimens from a large collection that had been brought together in its Washington, D.C., greenhouses for a taxonomic study of the family. Frank F. Meyer, later of the Federal Office of Plant Introduction, sent him a spineless opuntia he had found growing in a garden in Mexico. In addition to all these, Burbank said, he had varieties from Maine, Iowa, Missouri, Colorado, California, Arizona, New Mexico, the Dakotas, Texas, and other states. It was a vast collection, one of the largest of his breeding projects ever.

Over a period of eight to twelve years, the Burbank principle of mass selection was applied—hybridizing, selecting, and hybridizing again. The aim of the breeding was twofold: forage and fruit (the prickly pear, a common item of diet in Mexico and Central America, as well as in some Mediterranean countries). Many varieties combined both objectives, and some had beautiful flowers as well. Burbank ultimately introduced a total of thirty-four forage varieties.

Burbank admitted that the cacti sent him by David Fairchild, among others, had been spineless "for all practical purposes." However, he insisted, "I have still to see any form of Opuntia that is of good size and suitable for forage and yet that is altogether free from spines and spicules, except the ones that have been developed on my experimental grounds, and their progeny; and no such variety has yet been reported, although the authorities of the Agricultural Department of Washington scoured the earth to find such a variety." Others, notably Dr. David Griffiths, senior horticulturist of the U.S. Bureau of Plant Industry, disagreed. "Spineless as applied to Opuntias is a relative term," said Griffiths. "I have never seen one entirely spineless and much less spiculeless." W. B. Alexander, of the Australian Institute of Science and Industry, asserted that "the 'spineless cactus' is merely a practically spineless form of the well-known Indian Fig, *Opuntia Ficus-indica*. . . . Mr. J. H. Maiden states that specimens obtained from Mr. Burbank are in no way different from plants which have been in the Sydney Botanic Gardens for very many years." It would be fair to assume that Burbank had improved his opuntias but not entirely succeeded in getting rid of their defensive armor. The character of spinelessness in cacti is variable, even when they are propagated vegetatively, and the ones he sent out, or which were distributed by others, were not necessarily identical with those he had raised himself.

The controversy as to just what he had, or had not, succeeded in doing with his opuntias was aggravated by the fact that when he commenced to advertise his product, it was in the usual extravagant nurseryman's manner. The Carnegie grant—intended at least in part to support this work—had been insufficient to finance his breeding and development program. He had sunk funds of his own into it. As a businessman, he saw no reason why he should not recover his costs by sales. George Shull has a story highly revealing of Burbank's attitude in this respect:

> Just inside his gate at his Santa Rosa experimental garden, he had planted a bed, some 15 feet square, with the sprawling, thorny cacti from the desert. In the midst of this forbidding-looking culture, he planted a single specimen of Opuntia Ficus-Indica of the spineless variety, in most striking contrast with the thorny cacti around it. Mr. Burbank's visitors, who often came in droves, would look over the fence at this striking demonstration and comment to one another [on] the amazing wizardry which "created" the smooth fat-slabbed cactus from the sprawly thorny ones. I was a bystander on one occasion when a group of representatives of the press stood inside the gate beside the cactus bed, accompanied by Mr. Burbank. The conversation among the members of the press was much the same as that of the *hoi polloi* on the outside. Mr. Burbank listened to their conversation in silence, but when one of the men asked him whether he had actually started with the thorny types to produce the spineless cactus, he quietly gave the monosyllabic answer "No", but vouchsafed no further explanation. This is only one of many examples of his ability to set up an exhibit which was misleading to the uncritical. Mr. Burbank's own explanation of this had some merit, as you will readily agree. He explained that he had given up a successful and fairly lucrative nursery business to devote his time to the creation of new and improved horticultural varieties. In order that his plant-breeding program might be self-sustaining it was necessary for him to secure all the free advertising he could. Consequently, he felt that it would be foolish for him to reply to any sort of fantastic statement about his work if it were couched in terms calculated to enhance his reputation as a successful producer of marvelous new things.

He took pleasure in rubbing a cactus slab against his cheek to show how harmless it was. But visitors were so frequent that he could hardly cut a fresh slab for every demonstration, and the same piece was used again and again, till it was polished smooth by repeated rubbings. The more observant among his visitors sometimes noticed that there were numerous thorns on other parts of the same plant. (He was prepared to suffer for his cactus project and remarked that he had been pricked so many times in the hands and face handling the slabs that he sometimes had to shave the spicules off with a razor or rub them down with sandpaper so that as what was left of them worked into the skin, they would not cause more than minor irritation.) These ploys should have fooled only the ignorant and the unwary, but the journalists who took up the spineless cactus story so enthusiastically seem to have had no trouble accepting Burbank's presentation at face value.

To the specialists of the Department of Agriculture, it smacked of yet another phony promotion, analogous to the olive and eucalyptus booms. Experiment stations were swamped with inquiries about the value of spineless cactus. Their replies—based on the official conviction that cactus was at best an emergency source of feed to fall back on in bad years—were distinctly negative. They pointed out that cactus had spines for a reason: spineless cacti planted unprotected in the desert would be eaten by rodents before they got a start, and older plants would be killed by grazing animals eating them to the ground. Moreover, if they were to make the rapid growth Burbank advertised, they would have to be irrigated, and in that case alfalfa and other crops, which would give equal or greater tonnage of more nutritious dry fodder, which could easily be handled by machinery, were to be preferred. There was a great deal of acrimony on both sides, and in the end the spineless cactus episode contributed substantially to damaging Burbank's already precarious reputation with horticultural officialdom.

But in 1906 the cactus debacle was largely still in the future. In January of that year, Burbank's confidence in his "new creation" was strengthened when his Australian customer John M. Rutland purchased five varieties, designated by name as Santa Rosa, Sonoma, California, Chico, and Fresno, for $3,000,[4] as part of an order, also including six varieties of plum and his Montecito grape, totaling $5,800. On the strength of this, it must have seemed likely that his cacti would prove even more profitable than the plums that had hitherto been his greatest success. So encouraged was he that he built himself a large new house, which he liked to say had been paid for "from the proceeds of a sale of spineless cactus to a dealer in Australia." He was fortunate in that the contract for this fourteen-room, two-story dwelling (which stipulated that work be begun within a hundred days and set the cost of construction at $4,485) was concluded a few days before the great San Francisco earthquake of April 1906, which also devastated Santa Rosa.

The earthquake actually served to enhance belief in Burbank's miraculous powers among the credulous. Much was made of the fact that though the old house only a few feet away sustained severe cracks and had its chimney knocked down, his greenhouse had gone unscathed, and no damage had been done to his plants. At the Carnegie Institution, President Woodward noted hopefully that "in one respect, doubtless, the earthquake was advantageous to him and his work, namely in preventing visitors from encroaching too freely on his time and attention."[5] Having committed themselves, the Carnegie people were eager to see their protégé justify their faith in him.

In May 1906 George Shull arrived in Santa Rosa accompanied by members of the committee of the Carnegie Institution's Division of Biology. It was agreed that Shull would spend part of each summer for several years observing Burbank's work, and he stayed on until the end of June that year to make a start.

Shull was then thirty-two, a quarter of a century Burbank's junior. An Ohioan with a doctorate from the University of Chicago, he was a staff member of the Carnegie Institution's Station for Experimental Evolution at Cold Spring Harbor, Long Island, established in June 1904 under the direction of C. B. Davenport. There he had "made a preliminary report that outlined a program consisting of the development of a herbarium, the production of hybrids, a search for the causes of variation and, in particular, research into the nature of some of the mutations of *Onagra* (= *Oenothera*) *lamarckiana* obtained from Professor de Vries in Holland."[6] Shull was to study the phenomenon of mutation over a thirty-eight-year period, both in the evening primrose that had drawn De Vries's attention to it in the first place and in the shepherd's purse (*Capsella bursa-pastoris*), a plant of the mustard family found to mutate just as freely. In the light of his subsequent career (see pp. 169–71), he must be considered one of the foremost American plant geneticists of the day. It is hard to imagine a man better qualified to observe Burbank's work, to which his own concerns were of substantial relevance.

Shull soon found, though, that there were difficulties here quite unlike those normally encountered by botanists. The study of the man himself, he rapidly came to realize, would be one of the key aspects of his task; for this, he noted, "must be the most important factor in any comprehensive account of his work."

At the outset, Burbank was ill with one of his frequent colds. It was several days before Shull was even able to talk to him about his mission. Very well, he conceded, he needed a day or two to get settled in himself. Then, however, though he was treated with consideration and everything necessary to the study of Burbank's collections was thrown open to him, he began to discover, in his own words, that "certain psychological traits of Mr. Burbank made impossible the adequate study of his processes while he is at his work, because chiefly of his great sensitiveness to the presence of another person,—the self-consciousness induced by such presence necessarily lessening the clearness of his ideals and the spontaneity of his methods." (When Shull speaks of Burbank's "ideals," he means, of course, the ideals he worked toward in selecting his plants.) It was a sort of Heisenbergian uncertainty principle—the observer influencing the observed so that accurate observation became impossible.

Nonetheless, Shull persevered. Almost every morning he was out in the gardens by six o'clock. This was necessary if he was to examine the cultures before the workmen, who now did a good deal of the less exact selection, had destroyed any specimens. Remaining to watch this selection on a number of mornings, he found it to be "a complex process which no two workmen carried out on exactly the same lines, the result being that much that was left was essentially equivalent to much that was discarded." It was impossible to make meaningful comparisons between discarded and saved material.

Burbank usually came out at about 8:00 A.M., and Shull was able to accompany him about his work. He was made aware that his presence hindered Burbank from working as rapidly as he would have done alone. Accordingly, as soon as he had observed all Burbank's methods, he avoided spending time with him while he was working.

All of Burbank's methods, Shull noted, were of the simplest kind. None of them would have been considered satisfactory by an investigator interested in the genetic relationships of the plants he worked with. Burbank himself admitted that had he been conducting a scientific experiment, he would have done things differently.

But, he insisted, his work was nonetheless scientific for all that. It was in fact of a *higher* scientific type, he maintained, since he achieved the desired results without superfluous operations or unnecessary expenditure of effort, thus permitting the accomplishment of more work than would otherwise have been possible. Shull discovered, as others had done before him, that Burbank keenly resented suggestions that he was not a scientific man.

Until about 1904, he told Shull, he had done all the breeding and selecting with his own hands. Now, as a result of the great volume of correspondence and the number of visitors he received, he was obliged to delegate much of the less particular work to his men. He instructed them in the common characteristics he did not want, and these were gradually eliminated morning by morning, leaving only the task of making the final selections to Burbank. Almost all his experiments were begun at the home tract in Santa Rosa, and those that required continuous attention were carried on there. On the larger experiment ground at Sebastopol were trees and plants that needed only occasional attention. Shull observed that during his own stay Burbank visited the Sebastopol grounds on an average of once a week.

In general, the methods used by Burbank consisted almost entirely in cross-fertilization and selection. In crossing two plants, he usually removed the anthers before they opened to prevent self-fertilization. The pollen was then applied either by brushing the stigmas with freshly opened anthers, by brushing about in one flower with a camel's hair brush and then using this to pollinate the other, or by collecting unopened anthers in a watch crystal or a tin box lid, drying them, and then applying the pollen to the stigma by finger or with a brush. Shull was pained to note that in many cases no attempt whatsoever was made to prevent self-fertilization and that steps were only infrequently taken to prevent crossing with any other flower that might grow in the vicinity. Visits by insects were occasionally prevented by tying together the corolla after pollinating, but this was more often than not to prevent the delicate stigma from desiccating rather than to guard against the entrance of foreign pollen. To make sure that it was free of pollen from other sources, Burbank would often examine the stigma with a magnifying glass before applying pollen to it. If a few grains were detected, he would simply blow them off with a sharp puff of breath. He would then smear the stigma thickly with pollen and take the chance that even should an insect visit it, the likelihood of obtaining a hybrid was strong.

Before pollinating, he usually removed all buds except for one, or a small cluster, marking the branch that bore it with a strip of cloth. When pollen from different sources was used on the same plant or batch of plants, he marked them with different colors of cloth, depending on his memory, for the rest, as to exactly what the cross had been. When a number of different crosses were made among several species, he cut tags into different shapes and assigned one shape to each species from which pollen was being used. One tag of each form was then labeled and kept in the office as an index. The others, attached to the tree, were left blank, and indicated the origin of the pollen by form alone. When material was scarce and he was especially anxious to obtain hybrids, he would actually use pollen from several species on the same flower. Such methods, observed Shull, in a preliminary report to the Carnegie Institution dated July 1906, "lead to results that can not give any confirmation of Mendelism or any other theory of inheritance that rests upon statistical inquiry."

Selection, Shull points out, was the most basic of Burbank's methods since irrespective of whether or not cross-fertilization were resorted to, it entered into the production of every new variety. As to mutations, Burbank admitted they took place but held that they differed from the ordinary variation within species only in degree and not in kind. He maintained that any variation could be "fixed" by repeated use as a seed plant in the pedigrees—though some might require more repetitions than others. He used hybridization more often than not simply to increase the range of variation rather than to achieve certain desired combinations of characters. His other methods of increasing variability were the provision of especially favorable soil conditions and the use of imported material. Shull listed four elements he felt went to make up Burbank's remarkable talent for selection:

1. Sensitivity to slight variations in any desirable direction, which made for speed and accuracy in selection. "He says that he used to utilize the delicate senses of ladies in making his selections with respect to color, grace of form, scent, etc., but does so no

Samuel Walton Burbank.
Courtesy, Luther Burbank Museum, Santa Rosa.

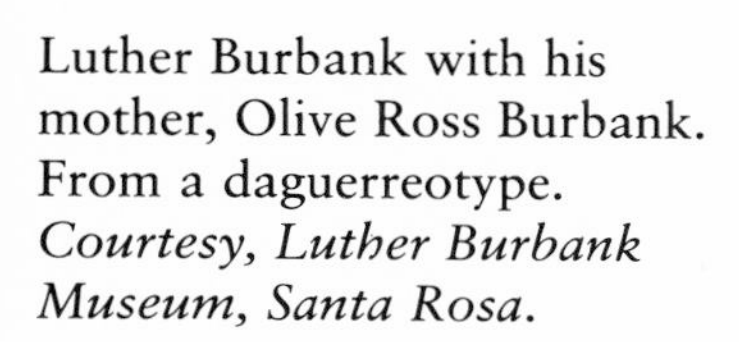

Luther Burbank with his mother, Olive Ross Burbank. From a daguerreotype.
Courtesy, Luther Burbank Museum, Santa Rosa.

Burbank in 1864, aged about fifteen. *Courtesy, Luther Burbank Museum, Santa Rosa.*

Burbank (right) with his sister Emma Burbank Beeson and, perhaps, her husband. *Courtesy, Luther Burbank Museum, Santa Rosa.*

Thomas Alva Edison, Burbank, and Henry Ford in 1915. *Courtesy, Luther Burbank Museum, Santa Rosa.*

Burbank, Hugo de Vries, and George Shull. *Courtesy, Luther Burbank Museum, Santa Rosa.*

"The Nestor of Our Horticulture": Edward James Wickson (1848–1923). *Courtesy, The Bancroft Library.*

Burbank with Jack London (right) and astronomer Edgar Lucien Larkin in 1908. *Courtesy, Oakland Public Library.*

Burbank with cactus. *Overland Monthly*, September 1908.

All Hail to Luther Burbank

By ADA KYLE LYNCH
INTRO
All Hail to Luth - er
Bur - bank, All Hail!
We greet you Luth - er Bur - bank, We
We love your fruits and flow - ers, We
(5) The peo - ple of all na - tions, Give
bow to your re - nown, Three cheers for Luth - er Bur - band, You won your Cap and
know you love them too, Your home's a beaut - eous bow - er, With flow'rs of ev - 'ry
hon - or to your name, In ev - 'ry land they know you, Your sto - ry and your
Gown, The wiz - ard of the flow - ers, They know your mag - ic
hue, (3) Great spears of Mol - ten Fire, Up - rear their heads of
fame, (6) The trees and fruits and flow - ers, Know well your lov - ing
wand, (1) In your own na - tive coun - try, (2)And ev - 'ry for - eign land. All
flame, (4)And all the grace - ful pop - pies, Add glo - ry to your fame. All
care, And give their buds and blos - soms, Their fruits and flow - ers fair. (7) All
Hail to Luth - er Bur - bank, All Hail!
Hail to Luth - er Bur - bank, All Hail!
Hail to Luth - er Bur - bank, All Hail!
INTERLUDE
Fine

"All Hail to Luther Burbank." Ada Kyle Lynch, *Luther Burbank, Plant Lover and Citizen; with Musical Numbers* (San Francisco: Harr Wagner, 1924).

Burbank with his second wife, Elizabeth Waters Burbank. "Photographed when he announced that he had presented to the world a new corn." *Courtesy, Oakland Public Library.*

Burbank crowned at the Santa Rosa Rose Festival in 1921. *Courtesy, Luther Burbank Museum, Santa Rosa.*

Burbank in his greenhouse with a Japanese pupil, Nobumi Hasegawa. October 1925. *Courtesy, Luther Burbank Museum, Santa Rosa.*

Burbank on stage near the end of his life. Seated behind the podium, immediately to his right, David Starr Jordan. *Courtesy, Luther Burbank Museum, Santa Rosa.*

longer because the long training has developed a higher susceptibility in him in these directions than is possessed even by ladies."

2. A quick and vivid imagination, able to visualize ideals by which selection could be guided. Sometimes a general ideal motivated him in a whole series of selections. In woody plants and fruit trees, for example, he saved only those with short, stocky internodes, prominent buds, and large leaves. In individual cases, his ideal was formed in response to suggestions given by the variation of the material itself. The suggestion for a blue Shirley poppy, for example, was the discovery of a decided bluish tinge in several specimens. The crimson Eschscholtzia was suggested by a delicate line of crimson inside one of the petals of an ordinary yellow one. "Sometimes his ideal grows as new possibilities appear in the material; but in the case of the Eschscholtzia it leaped at once to the perfect crimson flower."
3. Persistency. His ideals had to remain unchanged, or in harmonious development, sometimes for many years, "as any inharmonious shifting of the ideal which guides selection would lose all that had been gained by the previous selection."
4. The concentration with which he was able to consider a large number of different qualities without reducing the speed of the operation. In selecting fruit, for example, he had to consider not only size, shape, color, texture, and flavor, but also earliness of ripening, productivity, speed of reproduction, resistance to disease, insects, and frost, and cooking, shipping, and keeping qualities.

"There can be little doubt," Shull wrote, "that Mr. Burbank's success has been most largely due to the correctness of his economic ideals." In other words, he had an exceptionally keen eye for what the market wanted. His operation was economical, moreover, in the utilization of space and time. Selections were made at as early a stage of development as possible. Blackberries and cacti, for example, were selected in the seed pan, so that fewer than 1 percent of the seedlings were ever actually transplanted to the garden. Hundreds of grafts were made on the same tree, enabling Burbank to raise the entire progeny of a cross to maturity only two years from the planting of the seed. There were disadvantages to this when the grafts on a tree were from various sources, as it frequently became impossible to determine the origin of any particular graft except by guesswork. These trees, with their dozens of different kinds of fruit and foliage, had the effect of amazing observers who did not realize how easily explicable they were. Effects of this kind inevitably tended to swell the myth of Burbank's miraculous powers, and there can be little doubt that he relished the astonishment of the wondering public. In fact, there was no particular significance to this method of multiple grafting: it simply saved him space.

Shull carefully looked into Burbank's system of record keeping and confirmed that it was haphazard in the extreme. He conceded, however, that given financial and other limitations "more records could only have been coupled with fewer economic results." Those records that did exist were for the most part simply rough notes on the history and location in the experiment grounds of various plants, intended solely as an aid to Burbank's memory, on which he relied heavily for the more exact details. Only in his fruit records were observations recorded in any sort of scientific manner.

These fruit records consisted of single loose sheets on which the most promising of the selections being made were described. The fruit was cut in half and a pencil tracing made of the half-section on one corner of the paper. Anything noteworthy about the pit or core was then usually sketched in. A working name was assigned to the variety, heading a brief description as to size, shape, color, flavor, fruitfulness, cooking, shipping and keeping qualities, and so on. If Burbank thought particularly highly of a variety, he would mark the sheet with one or more double crosses indicating his assessment of its worth. If, after several years' testing, he decided that the fruit was fit to be put on the market, he would stamp "BURBANK" on the sheet with a rubber stamp as the mark of his imprimatur.

The whole system had more in common with the work habits of an artist than with any scientific procedure. And an artist, it may confidently be asserted, was exactly what he was, following his way among his living materials in almost complete reliance on his instincts and intuitions. "His whole life," Shull observed, "has been one of constant training of senses which were undoubtedly

1915 Older Plum

"Blue Point"

Near Chili Abutilon row East

Enormous size

Curious form

Black prune color

heavy blue bloom. flesh yellow deeply shaded scarlet and crimson firm rich delicious. ripe Sept 1st

should be a wonderful seller looks like a gigantic prune. half cling.

keyed to a high degree of sensitiveness by nature. Preconceptions do evidently affect his observations but principally in making him see the thing he is looking for to the partial exclusion of other things in which he has no interest, and not leading to any notable degree of error in the observations made." Moods tended to affect his observations to the extent of his noticing different features in the company of different visitors. For this reason, quite irrespective of any other, he sought solitude when he was at work. In the process of selection, the weighing up of the multitudinous factors involved was to a great extent done subconsciously, by intuition, "amounting in its highest development to slight autohypnosis, in which condition the two factors, observation and imagination, are so intimately associated that neither he nor any one else could successfully draw the line between them, and this is doubtless the source of greatest error in his observations."

So vivid were the impressions he received from his plants, so graphically did his remarkable memory reproduce them, that it was impossible to determine in what proportions observation and imagination had been mingled in the original critical experience. Shull thought it probable that the more strongly subjective element of imagination frequently came to dominate in memory, with the result that Burbank's accounts did not always square with the observable reality. Even so, he conceded, "the main features of the original observation are probably reproduced with unusual accuracy." He had only rarely, he admitted, found a man more honest to his experiences than Burbank.

Artistlike, Burbank preferred not to be influenced by others in his perceptions as to the relationships of the forms he was working on. He allowed that he had his own preconceptions and strove to avoid these, Shull wrote, "with what he believes is success, but which appears to me to lack much of it." One of these preconceptions was the belief that "nothing but acquired characters may be inherited." This caused him to attribute variations to environmental causes to a degree not warranted by observation. To Burbank everything in heredity was explicable as a mingling of forces. He believed that there was purpose in nature, and this, Shull observed, led him to "assign to many characters a use that present knowledge discredits." Instinctively, he took sides—against Mendel, Weismann, and De Vries, and with Darwin and the contemporary opponents of Mendelism and mutation theory. In doing so, he generally did not trouble to find out exactly what it was that these men had taught.

In later years, it would be claimed that Burbank had independently discovered Mendel's laws. For example, in the supposedly autobiographical *Partner of Nature*, put together by Wilbur Hall, he is made to say:

> I slowly came to a realization of the fact . . . that all the best variations and recombinations in a hybrid stock, obtained by crossing, appear in the second and a few subsequent generations—almost never in the first. This was, of course, Mendel's Law, that I had observed in many experiments long before Dr. de Vries and others unearthed the Mendel reports and made them belatedly public.[7]

And Shull believed that Burbank had sensed early in his career—probably before the rediscovery of Mendel in 1900—that the "break" in hybrid lines comes only in F_2 and subsequently, and never, or at least not usually, in F_1. But neither this nor the observation of ratios in the offspring of hybrids—which had been noted by Darwin and others long before—is the essence of Mendelism, the revolutionary element in which is the idea of segregating alleles and of dominance and recessiveness in the factors of heredity.

Burbank was, for that matter, inclined to dismiss or deny the ratios. Shull notes that on one occasion Burbank drew his attention to a row of seedlings of a hybrid between a purple-leaved and a green-leaved plum, in which a 3:1 ratio was to be expected in Mendelian terms, but which gave the impression of being about 90 percent purple. This was done with the specific intention of discounting the general validity of Mendel's law. When the younger man took the trouble to make an actual count, however, he found an almost perfect 75 to 25 percent relationship between the seedlings of the two colors. The superficial effect of an overwhelming preponderance of the purple-leaved type was an optical illusion.[8] Once, when he pointed out a discrepancy between certain published figures and the true numbers involved, Burbank told him "quite positively" that "I never count anything." And, Shull observes, no one who does not count can discover the Mendelian principle.

Figures presented one of Burbank's real weaknesses, so much

so that Shull estimated that those given in published statements should be divided by at least ten to get a true approximation of the size of his cultures. On one occasion he told Shull—and the press—that he had planted 4,000 *varieties* of seedling potato out of 14,000 to be planted. Shull counted them and found that there were in fact 4,000 *individuals*, belonging to 390 clones. He did not think this was dishonesty on Burbank's part, rather that, "the numbers simply so much exceed his capacity for numerical conceptions that any large number will seem to him to be reasonably moderate."

Paradoxically almost, honesty struck Shull as one of Burbank's consciously strong characteristics, the product of his powerful sense of self-esteem. He was of the kind, Shull thought, who would rather be right than be president. Though Burbank was inclined to speculate on subjects he knew little about, delivering himself of trenchant generalities unsupported by fact, he was otherwise frankly prepared to admit that he did not know something, even "in cases where it might appear to his advantage to know, and where his imagination could quite readily supply the information without the possibility of detection if he so desired."

Conflicting with the independence conferred by his self-esteem was his love of approval by others. Though he would do nothing dishonest to earn such approval (for that would have brought self-condemnation), he eagerly accepted it as no more than his due. "There are striking instances," says Shull, "in which the combination of these two dominant traits produces one instant the most profound modesty and the next instant almost blatant self-praise."

In 1906 he told Shull that only Darwin had grown more plants than he had and that he was thus the greatest authority on plant life next to Darwin. Two years later, "he figured that he had by that time surpassed Darwin in the number of plants he had raised and that *therefore* he was the greatest authority on plant life that had ever lived."[9] This being the case, he felt that he was better qualified than anyone else to pronounce on the subject of evolution.

While not claiming the powers of a wizard, Burbank was pleased that the public should think him one. He delighted in the role of benefactor of mankind. This feeling of superiority naturally had the effect of making him less susceptible to evidence from outside sources that might tend to contradict what he himself had deduced from experience. "He says he read very little of De Vries' 'Species and Varieties' because the author was working along such wrong lines; and he is always impatient of a conversation in which he does not do all or nearly all the talking," Shull observes.

Though his phraseology constantly stressed his own role as "creator" of his products (he "originated" and "trained" his plants, "teaching" them to "behave" in the manner desired), he would nonetheless give a straightforward answer to an informed question from a scientific interlocutor. Shull found that he could depend on sober answers, even though he noticed that other people sometimes received evasive ones. His appreciation of Burbank's love of being intelligently questioned about his work, he felt, materially aided their relations.

It is a psychological portrait of some acuteness that Shull presents: evidently his powers of observation and evaluation in that respect were by no means inferior to his horticultural sense. At once perspicacious and sympathetic, he was able to look Burbank's faults in the face without losing sight of his merits. He made no fetish out of objectivity, but neither was he overcome by the Burbank charisma. It is to the credit of both men that they managed to remain friends of a sort.

A more homely picture of Burbank is painted by W. C. Williams, an osteopath who attended him from early in 1906 to mid-1908.[10] Williams was living in Healdsburg, about sixteen miles from Santa Rosa, when he received a telegram summoning him to Burbank's house on a sick call. When he arrived, he was told that Burbank had pneumonia but, after examining him, diagnosed the problem as acute bronchitis. He visited Burbank daily for nearly a week and was finally persuaded to transfer his practice to Santa Rosa, Burbank telling him that he wanted a course of treatments himself and assuring him that "with his quietly talking to his close friends he would be able to build me up a nice practice what with Santa Rosa being a large town and all." In the course of the treatments, which took place in the evenings in Burbank's bedroom, Williams noted in 1948:

> He told what a struggle he had had to keep up his work on account of being short of capital. He told me that Carneiga [*sic*] Foundation was

> figuring on assisting him. Then he told me how trying it was on him to have Carneiga men there going all over his business and asking so many private questions. He said, "They seem to think because they aim to grant me money in case I qualify that they own my place and can come and go as they please." He told me that if he didn't need the money so desperately he would call the whole thing off and forbid them coming on his grounds.

Williams felt that the treatments, including a good deal of massage, did Burbank a lot of good, "because his circulation was poor and he was going through such a strain at the time." For his part, Burbank seemed pleased, telling him that "his experience had been with osteopaths that they would give you a good long treatment the first two or three times and from then on give you a short treatment and he liked a long treatment." He declared his belief in "natural methods" in treating diseases of the human body, saying that he had been one of the first to recognize osteopathy.

One thing that emerges clearly is Burbank's preoccupation with his health. "He told me he had never been strong in his life, always frail. Then his dry humor asserted itself, 'I never had a constitution—I always went by the bylaws.'" A surgeon had once bungled a hemorrhoidal operation, he said, "and it had been a great handicap to him ever since." They spoke a good deal about diet. Onions, Burbank insisted, "were healthful. Said he never knew a person that eat lots of onions but what was healthful." He was no vegetarian, however, and "told me he never saw a vegetarian but what he could take him by the seat of the pants and throw him over a fence."

Though Williams felt they were good friends, he found he had to keep his distance on anything pertaining to Burbank's private affairs. Questions were only rarely allowed. But Burbank did make a number of remarks that are of considerable psychological interest. "He told me how things would come to him at times," Williams recalled. "How ideas would come in his mind almost as if he had been spoken to." Jack London used to come over from the nearby Valley of the Moon in those days, and Williams remembered how much Burbank seemed to enjoy his visits. One day he had asked Burbank "what he thought of Jack London plagerizing [*sic*] in writing a book." Burbank replied tartly: "If someone takes an old abandoned house that no one cares to live in and remodels it and makes a fine house of it, is that wrong?"

It was his pragmatic approach in a nutshell. One might as well have said, "If Burbank imports a plant from Chile or Manchuria that no one has bothered with, gives it a name and makes it a useful economic crop, is that wrong?"

"I saw the point," Williams admits.

12

BURBANKITIS

Burbank, who was so much written about, had written little himself. Aside from his catalogues, he was the author only of letters and brief items in agricultural and horticultural journals and of one or two short papers. He preferred to get on with his work and let others do the theorizing. In May 1906, however, he published a lengthy article in *Century* magazine, which was reissued the following year as a small book. Curiously enough, it had nothing to do with Burbank's area of special competence. Instead, it was an essay on human eugenics and child rearing entitled *The Training of the Human Plant*. (In view of its interest as an illustration both of Burbank's views on heredity and of his relationship to the contemporary eugenics movement, this is reproduced in full as Appendix II.)

The term *eugenics* was coined by Charles Darwin's cousin Francis Galton (1822–1911), a brilliant statistician who was one of the first to attempt to understand heredity in populational terms. In *Hereditary Genius* (1869) and other books, Galton argued vigorously that intelligence and talent are inherited characteristics. "If a twentieth part of the cost and pains were spent in measures for the improvement of the human race that is spent on the improvement of the breed of horses and cattle, what a galaxy of genius might we not create," he suggested.[1] Galton rejected the idea that acquired characteristics might be inherited—which would make possible improvement by means other than breeding and selection—and put forward theories of inheritance that to a remarkable extent prefigured those of August Weismann.[2]

Rejection by a majority of biologists of the concept of the inheritance of acquired characteristics set the stage for the development of an organized eugenics movement. This took place after the turn of the century in both England and the United States. In America, says Richard Hofstadter, the eugenics movement grew so rapidly that "by 1915 it had reached the dimensions of a fad."[3]

In January 1906, the American Breeders' Association, of which Burbank had been made an honorary member, set up a Committee on Eugenics with a mandate "to investigate and report on heredity in the human race" and "to emphasize the value of superior blood and the menace to society of inferior blood." The membership of the committee amounted to "a *Who's Who* of the founders of the eugenics movement in America," says Mark Haller:

> David Starr Jordan, its chairman, a prominent biologist and Chancellor of Stanford University; Alexander Graham Bell, famous with the public as an inventor but famous among eugenists for his sheep breeding and his study of hereditary deafness; Charles R. Henderson, a University of Chicago sociologist, well known in all reform circles and a respected student of crime and poverty; and Luther Burbank, who remained something of an anomaly in the movement because he retained a belief in the inheritability of acquired characteristics. Other members included Roswell H. Johnson, geologist and later a professor of eugenics; Vernon L. Kellogg, a biologist at Stanford University; and William E. Castle, the Harvard geneticist. But the most important member was Charles Benedict Davenport, whose name dominated, for good or ill, the early eugenics movement.[4]

Davenport was Shull's chief and the Carnegie Institution's director of experimental evolution at Cold Spring Harbor. His views on the subject of eugenics were deeply held and tended to be expressed in inflamatory language. "Society must protect itself . . . as it claims the right to deprive the murderer of his life, so also it may annihilate the hideous serpent of hopelessly vicious protoplasm," he declared.[5] Though a kindly man and not overtly a racist,

> he failed to cast aside the presuppositions of the overzealous Mendelian. He became firmly committed to the notion that almost all traits in man, physical and mental, are inherited in simple Mendelian patterns, and he paid only token tribute to the role of environment in determining a person's makeup. Much of his later research consisted of attempts to demonstrate that such characteristics as wanderlust, musical ability, inventiveness, alcoholism, laziness, truancy, and jealousy are determined by single genes. The fact that a trait did not skip a generation in a family was sufficient proof to him that it was a Mendelian dominant.[6]

As Haller puts it, Burbank's association with the eugenics movement was "something of an anomaly." Not only did Burbank argue that acquired characteristics are inherited; he is visibly groping in *The Training of the Human Plant* for a theory of inheritance remote from the determinism of contemporary biology, with its divisive belief in the permanent allotment of "the gifts of potentiality" by immutable "unit characters of the germ-plasm."[7] Burbank was never one for orthodoxy. He always tried to think things through for himself. "Heredity means much, but what is heredity?" he asks.

> Not some hideous ancestral specter forever crossing the path of a human being. Heredity is simply the sum of all the effects of all the environments of all past generations on the responsive, ever-moving life forces. . . .
>
> . . . environment is the architect of heredity; . . . acquired characters *are* transmitted . . . *all* characters which *are* transmitted have been acquired, not necessarily at once in a dynamic or visible form, but as an increasing latent force ready to appear as a tangible character when by long-continued natural or artificial repetition any specific tendency has become inherent, inbred, or "fixed," as we call it.[8]

However one may judge this from a scientific standpoint, it cannot be equated with the stance of Davenport and his fellow believers. Indeed there was so little in common between Davenport's views on heredity and Burbank's that it is hard to see the latter as an active member of a Committee on Eugenics of which Davenport was the moving force. Burbank's nomination to serve on the committee was probably as much an honorific gesture, in fact, as his election to membership of the American Breeders' Association had been in the first place. In it we may perhaps detect the influence of Burbank's well-meaning friends at Stanford, David Starr Jordan and Vernon Lyman Kellogg. To Burbank himself, acceptance may have seemed no more than a matter of good public relations—painless association with a distinguished group of men proclaiming high scientific and humanitarian ideals.

In conceding in *The Training of the Human Plant* that it would be best to prohibit the marriage of the "physically, mentally, and morally unfit," Burbank fell into line not only with contemporary biological dogma, but with public opinion. A law providing for the sterilization of the mentally incompetent was passed in Indiana in 1907. Two years later, California became the second state to pass legislation of this kind, and by 1915, sterilization laws were in force in twelve states. Twenty-nine states had such laws by 1940, and over 33,000 people had been sterilized nationwide.[9] By then, however, it had already become apparent to geneticists that feeblemindedness is not simply a matter of a single recessive gene, and that eugenic programs were unlikely to be effective even if it were.[10] More important, perhaps, revulsion at the applications of a distorted version of eugenicism by the Nazis in Europe had begun to deprive the movement of such support as it still enjoyed in America.

It is deceptively easy in hindsight to damn the early eugenicists as racists. "The blood of a nation determines its history. . . . The history of a nation determines its blood," wrote David Starr Jordan.[11] Stripped of its context, such an assertion conjures up emotions today that would have astonished and horrified its author. Far from being a bigot in pursuit of racial supremacy—white hegemony was, in any case, a simple fact of life at the time—Jordan was a deeply committed pacifist taking part in what he perceived as a life-and-death struggle against militarism, which not many years later was to rip apart the fabric of Western society in World War I. It was an age when generals wrote Social Darwinist treatises arguing the beneficial effects of war in promoting the survival of the fittest.[12] "This doctrine has no legitimate connection with Darwinism," Jordan declared.[13] For him, the essential feature of war was "the slaughter of the young, the brave, the ambitious, the hopeful, leaving the weak, the sickly, the discouraged to perpetuate the race."[14]

Often misleadingly depicted as an offshoot of Social Darwinism, eugenics was actually in an important sense an appeal to the latest findings of science *against* Social Darwinist crudities that were proffered in justification of inhumane social policies at home and imperialism abroad. The brazen voice of Social Darwinist laissez-faire is now no more than an echo, and it is worth recalling that in 1905 Justice Oliver Wendell Holmes, Jr., found it needful (in connection with a case involving the right of the state to limit working hours) to point out to his colleagues on the Supreme Court that "the Fourteenth Amendment does not enact Mr. Herbert Spencer's *Social Statics*."[15] His was the minority opinion.

The eugenicists saw themselves primarily as humanitarian reformers. But the science to which they appealed—early Mendelism and Weismann's doctrine of the inviolable germ plasm—could all too easily be interpreted as providing a rationale for what we now see as racism. Notwithstanding Davenport's call for "additional precise data as to the unit characteristics of man and their methods of inheritance,"[16] eugenics dovetailed rather too comfortably with ordinary bigotry. Reason appeared to endorse prejudices that ran deep in American society. Only a saint with a preternatural grasp of future trends in genetics could have stood out unambiguously against the combination.

Luther Burbank was neither a saint nor a geneticist. He was a horticulturist working in a small California town and not exempt from the biases of his contemporaries. In the latter part of the nineteenth century, with Denis Kearney's "Sand-Lotters" and other populist agitators whipping up sentiment in San Francisco, the West Coast was swept by hostility to Chinese immigration. It was felt that large numbers of Chinese immigrants simply could not be incorporated into American society. A small Chinatown had sprung up in Santa Rosa, and in 1886 we find a press report listing Burbank as a member of a committee of local exclusionists. In 1892, moreover, Burbank wrote a congratulatory letter to a local congressman, Thomas J. Geary, in connection with the latter's efforts to extend the federal Exclusion Act of 1882, which prohibited Chinese immigration and naturalization. (It may be argued in Burbank's behalf, however, that he seems to have done so chiefly as a device for soliciting Geary's backing for a plant patent law, the subject to which most of the letter is devoted: "As a nation horticulturally speaking we are not less than fifty years behind what we should have been if the granting of patents had been allowed to plant originators," he wrote. "Why should they not be?")[17]

In the final analysis, *The Training of the Human Plant* is the thesis of a deeply humane man struggling to merge what he knows with what he feels. To the end of his life, Burbank periodically voiced—or permitted his ghost-writers to voice—eugenic opinions that ran counter to his own belief in the shaping potential of the environment.[18] But though his ideas may have been contradictory, there is no doubt as to where his heart lay. This is evident from his passionate opposition to what he perceived as the regimentation and corruption of children by the educational system. "No boy or girl should see the inside of a school-house until at least ten years old," he insists. Rather, children should spend their formative years in the open air, in close contact with nature. They should be educated with love and honesty, not by compulsion and deceit. Foreshadowing Ivan Illich's *Deschooling Society* (1971), Burbank goes so far as to propose that we "break up this cruel educational articulation which connects the child in the kindergarten with the graduate of the university." As it is, he says, "The work of breaking down the nervous systems of the children of the United States is now well under way."[19] This was, and remains, a direct challenge to one of the greatest American shibboleths. But the educationalists did not seem to notice—just as the eugenicists failed to notice that Burbank was out of place in their company.

Ella Wheeler Wilcox ecstatically proposed that *The Training of the Human Plant* be reprinted by the government and a copy sent to every woman in the country. Every mother and mother-to-be should read it, she declared. It ought to be preached from pulpits, discussed at women's clubs, and taught to young women in seminaries and colleges. The world had adopted Luther Burbank, and it was determined to understand him in its own way. He was beginning to be a legend.

Burbank had early on subscribed to a newspaper clipping service. Preserved in scrapbooks, these clippings are a clear measure of his growing celebrity. The magazine articles about him by Wickson, Harwood, Jordan, Kellogg, De Vries, and others, Harwood's book *New Creations in Plant Life*, and the chapter devoted to

Burbank in Harwood's *The New Earth* served as the basis for countless newspaper articles. For many years, reporters had been eagerly seizing on any statement Burbank cared to make in their search for "copy," further swelling the flood of newspaper publicity. He was usually only too willing to provide them with quotable material. ("I don't believe I ever called on him that he did not give me a good story for my paper," one told Howard.) All these items were clipped to be filed away by the faithful Emma in large quarto volumes kept for the purpose.

The earliest clippings, dating from 1874, were reports from the Fitchburg *Sentinel* and various other local papers about the Burbank potato. By 1906 the seventh volume of the scrapbook was in the process of being filled, and clippings were accumulating at the rate of some 600 quarto pages a year. From each original source, the flood of publicity spread out in concentric circles. Volume 1, covering the period from 1874 to 1895, had 150 pages. Volume 2, running from 1896 to 1901, had 204. Volumes 3 and 4, dating from 1901 to 1905, made up a combined total of 400.

Honors were showered on Burbank. In December 1903, he was made an honorary member of the elite Bohemian Club, a distinction rarely conferred (other honorary members of the day included Mark Twain and Theodore Roosevelt). In 1905 he was awarded the honorary degree of doctor of science by Tufts College (now Tufts University) as part of the celebration of its fiftieth anniversary. That year, too, he was made an honorary member of the California Academy of Sciences, and a testimonial banquet was given for him by the California State Board of Trade. In 1906 the California Federation of Women's Clubs passed a resolution recommending that his birthday, March 7, be celebrated as State Arbor Day. The following year, Santa Rosa bestowed his name on its newest grade school. (In the end, some twenty schools were to be named after him in California alone.) He declined the distinction of having a brand of cigars named after him.

Meanwhile, however, some in the profession of horticulture were becoming increasingly resentful of his swelling reputation. Other plant breeders, it was felt, had done just as much, or more, but received nowhere near the rewards that accrued to Burbank. "I am heartily tired of this gush about Mr. Burbank," wrote Ernest Braunton, secretary of the Southern California Horticultural Society. He held nothing against the man himself, he said. "I am not of that envious class of tradesman who ask, 'What has he done,' but really consider him the most overestimated man, by the general public, that we have in the world today." Braunton observed that he himself did a good deal of breeding, particularly of gladioli, of which he had some that were superior in size and color to anything he had seen. He failed to see that Burbank had accomplished anything that he and many others could not have done—or did in fact do—in the same time, and he vowed "to do what I can to get the public to look at him through normal eyes." As for the trade, they were so "touchy" that "the best of them rise up in arms at the mere mention of his name." Members of the Southern California Horticultural Society had opposed inviting Burbank to talk, even though, Braunton observes, "I know it would be the best drawing card with the public that the Society could have."[20]

Even the faithful were feeling the effects of the condition now dubbed "Burbankitis." Horticulture, the loyal Wickson observed, replying to Braunton, was suffering through the exaggerated reports published by ignorant people about Burbank's work. The latter was himself, perhaps, the chief sufferer. "I have done what I could to stem this tide of exaggeration and false conclusion," Wickson says, "but have become somewhat weary of the task, and have rather adopted the policy of waiting until the misrepresentation and exaggeration shall expire by its own unfitness to live. I doubt if there is any use in continually writing against this misrepresentation, and for this reason I have largely retired from the field."[21]

Now Wickson, that "Nestor of our horticulture," was one of the chief culprits himself. It could perfectly well be argued that *he* was the one who had convinced the public of Burbank's unique talents and contributions to horticulture in the first place. No other writer of scientific standing had worked at promoting Burbank with anywhere near his enthusiasm and efficiency. He had in the past been guilty of the most astonishing hyperbole on Burbank's behalf.

His cooling off can perhaps be explained in terms of his mounting academic prospects. In 1905 he had been elected dean of the College of Agriculture of the University of California and appointed acting director of its Agricultural Experiment Station, stepping into the shoes of the great E. W. Hilgard. The following year, Wickson became a member of the National Horticultural Council. Full professorship at the College of Agriculture and appointment as permanent director and horticulturist of the Agricultural Experiment Station were in the offing and would come his way in 1907. In the meantime, it doubtless behooved him to be more cautious about his endorsements than he had been in the past.

A warning note was sounded by Patrick O'Mara, horticulturist for Peter Henderson and Company, a New York seed and nursery firm that had bought some of Burbank's productions and listed them in its catalogues, who had visited Burbank in the summer of 1906. Though this was the busiest time of the year for the plant breeder, O'Mara, who was in charge of the plant department of one of America's leading seed houses, expected a cordial reception, to say the least. For some reason, however, he was totally snubbed. Not only did Burbank refuse to see him, he would not even admit him to his experiment grounds.

This would probably have incensed almost anybody. His pride wounded, O'Mara retaliated by attacking Burbank at a meeting of the New York Florist's Club and followed up his initial salvoes with a sarcastic article entitled "Luther Burbank: A Short Review of His Work in Plant Hybridization and Brief Comparison with Other Hybridizers," which appeared in the *Florist's Exchange*, a trade publication.

O'Mara began by taking Burbank to task for his claim to "new creations." He had, O'Mara said, "no more right to claim the title of 'creator' of new plants than he has to apply it to the bee that flits from flower to flower and carries the pollen." He was particularly scornful of Wickson's set of articles in *Sunset* magazine. "Many friends and admirers of Mr. Burbank contend that he is not responsible for the extravagant claims made for him in that publication; but it is well to bear in mind that he helped to circulate it and therefore gave a semblance of sanction to its contents." Certainly, the critic did not have to search very hard to find passages that were eulogistic to the point of parody:

> Plant development is one of the phases of civilization, and it makes new conquests as they are needed in the onward rush of mankind. We

> are now at the beginning of an epoch of accelerated motion in this direction. Burbank is the prophet of this epoch. Obeying the command of the Infinite, he is carrying the gates of Gaza. Let not the Delilah of modern organization shear him of his God-given strength and make him like other men.[22]

And so on. Adulation in such terms was almost as bad as slander. The hero would have had to be a Samson indeed to bear up to it. "Then," O'Mara appended rancorously, "he reaches out and gets the $100,000." (This presumably refers to the Carnegie grant of $10,000 a year for a projected ten years, the second installment of which had been paid in January 1906.) There was more than a tinge of plain envy in the attack. In addition to being heaped with honors and awards, Burbank was making more money than the average horticulturist could ever hope for, and getting subsidized by Carnegie into the bargain.

Why had not other deserving men had similar rewards? O'Mara ventured that the loganberry, a hybrid between the wild California black dewberry and the red raspberry, originated by Judge J. H. Logan in the early eighties, was better than any of Burbank's productions. He exclaimed over the fact that "the Phillips cling peach is of more value to California in my opinion than any fruit which he has produced, and sad to relate, the man who produced it is in the Yuba Co. almshouse, so announced in the Pacific Rural Press of Jan. 6th last." He cited hybrid grapes developed by Rogers, Jacob Moore, and T. V. Munson, the work of John Cook, E. G. Hill, and Dr. W. van Fleet with roses, that of Fred Dorner, C. W. Ward, and Peter Fisher with carnations, Antoine Wintzer's work with cannas, and Groff's with gladioli, inter alia, as evidence that Burbank was far from being the solitary pioneer he was cracked up to be.

He dismissed Burbank's science as "somewhat of the Mary Baker Eddy or Helen Wilmans order." As to all the talk about "educating" plants, "I wish to say, and make it as emphatic as possible, that in my opinion no exercise of the human mind by way of suggestive thought directed upon a plant can change one cell or filament of it." This last is interesting. For though Burbank never wrote that he influenced plants by "suggestive thought"—or indeed permitted himself to be quoted to that effect until much later—it is evidence that rumors of the kind were current. Given the backdrop of the "Messiahship idea" and his mystical propensities in general, it might not be too much to assume that such rumors originated with Burbank himself. Whether or not he believed it (and he probably did), it was exactly the sort of power he was likely to hint at to a gullible audience—and such he had aplenty. The reference here to Christian Scientist Mary Baker Eddy (1821–1910) is also revealing: there was certainly something approaching evangelical fervor to "Burbankitis."

O'Mara was not of a disposition to allow his subject much credit for anything. The Burbank potato, he alleged, was a volunteer seedling of the Early Rose, not hybridized by him. In the East, he contended, it had outlived its usefulness and been superseded. He was rude about the Shasta daisy, branded the Burbank plum "an importation pure and simple," and observed that Carter of London had produced a superior crimson Eschscholtzia* and Italian breeders had developed better cannas. As to the Iceberg blackberry, he said, accurately enough, "Nonsense, people don't want a white blackberry; they want a black blackberry."

Needless to say, he took off after the current miracle—the spineless cactus. First of all, the original plant "was given him right straight out of the Department of Agriculture." Secondly, what was it good for anyhow? "I never crossed the desert myself except on a railroad train," O'Mara observes with elephantine humor, "but I can imagine that if a man is crossing the desert and wants to sit down, how handy it would be to have one of these thornless cacti handy." Here again the hyperbole of Burbank's admirers could be used against him. "They say when a poet gets after a man, he is done for," O'Mara gibed, turning up a deadly verse in evidence:

> He touched the spiculed desert—cacti cursed—
> And turned its thorns to figs; its thistles fruit.
> He nodded to the daisy half immersed
> In dwarfing dust, and lo! a lily mute
> Rose from the weeds—a perfume with a flute.

To this awful celebration, he then appended a bit of comic doggerel, supposedly spoken by one of the "knockers":

* Several English breeders also produced "crimson" Eschscholtzias. Carter and Company's "Rose Cardinal" and Watkins & Simpson's "Intus Rosea" were both introduced in the early eighties, Shull notes.

O, Mr. Burbank, won't you try and do some things for me?
A wizard clever as you are can do them easily.
A man who turns a cactus plant into a feather bed
Should have no trouble putting brains into a cabbage head.[23]

The haw-haws in the Florist's Club must have been resounding. But if some of this sniping was justified by a few grains of truth, it was not for that reason any the less unfair. Certainly Burbank had imported; certainly he had received cacti from the Department of Agriculture. But his importations had been subjected to elaborate crossing, and his opuntias had been grown and selected on a vast scale before being put on the market. It was true that other men had done good things in various fields. But no one had attempted a fraction of what he had. A more valid criticism would have been that he had tried to do too much.

Burbank found his defenders, notably W. Atlee Burpee, the famous Philadelphia seedsman, who had long been appreciative of his work. (Burpee, O'Mara observed, tried "to show me the light, but I could not see the light.") But the run of the trade was henceforth increasingly suspicious, if not openly hostile.

Every great reputation provokes detractors, and Burbank's called forth enemies in greater numbers than most precisely because it was so much bolstered by addle-brained sentimentality and exploited with increasingly incautious puffery. The claque of hack writers, mystics, and schoolmarms who sang his praises so loudly inevitably raised the hackles of a good part of the nursery trade, which had reason to be jealous anyway, and of the scientific community as represented by the workers in the agricultural experiment stations and the staffs of the land-grant colleges—who were, it was touted abroad, being put to shame by this unlettered man. Once their enmity was aroused, it was uphill work to convince them that Burbank was—despite all that—doing remarkable things. They would not see the light. The dog had been given a bad name. Hang him.

O'Mara's attack met a warm enough reception to warrant its republication in pamphlet form, which the author did in Jersey City in September of the following year. Conversely, however, 1907 also saw publication of Hugo de Vries's book *Plant Breeding: Comments on the Experiments of Nilsson and Burbank*. A large section was devoted to Burbank's work, which the Dutch botanist clearly viewed as important. This was a significant compliment and was bound to be influential, particularly in Europe. Again on the debit side, Harwood's rhapsodic *New Creations* now went into a second edition, and he published yet another eulogy, entitled "How Luther Burbank Creates New Flowers," in the *Ladies Home Journal*.

But whatever the current state of Burbank's fame, business was booming. There were candidates aplenty for the bandwagon. A list compiled by Emma Beeson records the sale, on September 20, 1907, of seven varieties of thornless cactus—Santa Rosa, Sonoma, California, Fresno, Monterey, Chico, and Guayaquil—to the Thornless Cactus Farming Company of Los Angeles for a grand total of $26,000. This was probably a speculative operation prompted by Burbank's twenty-eight-page illustrated catalogue "The New Agricultural-Horticultural Opuntias: Plant Creations for Arid Regions." Appearing in June, this inevitably whetted the appetites of the get-rich-quick brigade, always thick on the ground in California. It may perhaps be doubted that Burbank was paid in full (other figures suggest that only $20,000 was involved). He may well have been prepared to grant a certain amount of credit in light of the size of the order. It was, nevertheless, the biggest single sale he would ever make. Very likely it was the biggest single sale in horticultural history up to that time.

Burbank's reputation had reached its perihelion along with his earnings. The first marks of decline, so evident to hindsight, were by no means obvious then, however. O'Mara's ranting and the mutterings of the trade could be written off as envy. Burbank had been written up at length by one of the most famous botanists in the world; the deluge of favorable publicity continued.

Early in 1906, Burbank had begun negotiations with Dugal Cree, president of the Cree Publishing Company of Minneapolis, for the preparation of a ten-volume account of his work intended for popular reading. This was disturbing news to the Carnegie Committee, which was anticipating the publication of its own definitive account. "It seems incredible that Mr. Burbank should enter into the arrangement referred to," Davenport wrote to Woodward. However, Shull advised in February 1907: "Mr. Burbank informed me that the Cree books can not be in the least conflict with our work. He says that the 10 volumes will not contain as much 'meat' as ten pages of the Carnegie work. It is to be made up mostly of illustrations, on thick paper with very brief statements in large type:—to be sold by subscription."

Mindful of the criticism of Harwood's book, the Cree Company had hired Dr. Leroy Abrams, assistant professor of systematic botany at Stanford University, to read its manuscript and stand sponsor for its scientific aspects. Abrams, Shull noted, "will be on the ground here until the work is completed, which is expected to be at least a year."

For Shull that year was marked by tragedy. In July 1906, he had married Ella Amanda Hollar, and returning to Santa Rosa in 1907, he brought his young wife with him. She was well on into pregnancy and perhaps ought not to have made the trip. The baby, a girl, did not survive its birth on May 5, and Ella herself died two days later. She was buried there—her husband would be buried there himself almost half a century later—and Shull returned to Cold Spring Harbor alone.

When he next visited Santa Rosa, early in October 1907, he found relations somewhat strained. Burbank was evidently getting impatient with the delay in publication of the institution's findings and wanted them to bring out at least an initial volume. Shull found it difficult to get to see him at all. "In the nine days I have been here," he reported,

> I have had only two hours of Mr. Burbank's time. He comes in each day at the appointed hour to say that he "Simply can not give me the time" and promises to do better later. In view of the urgency on the part of the Committee to hasten the work, you can imagine the strain this puts on me. I am *feeling* it, too. I am not idle of course, for I am pushing the matter of collecting statements regarding Burbank's work from authoritative sources and have also brought with me Station work to which I can devote what time I am not otherwise occupied. I can not avoid the feeling sometimes, however, that I would be accomplishing much more for science if I were back with you and would not then be taking Mr. Burbank from work he is so preeminently fitted to do.

Shull himself could scarcely have appreciated the full irony of the situation, for the work he was doing at Cold Spring Harbor with purebred lines of corn was to lay the basis for the most important development in agriculture in modern times.

Shull had not started out with the idea of improving corn. "His objective was to analyse the inheritance of quantitative characters and he chose corn as an appropriate experimental subject for this purpose," the geneticist Paul C. Mangelsdorf observes. "He inbred his lines to 'fix' their characteristics and he crossed lines which had thus been inbred to study the inheritance of kernel-row number."[24] The resulting hybrids dramatically demonstrated the effects of the phenomenon Shull called "heterosis"—otherwise known as hybrid vigor. They were superior to the original strains both in vigor and in productivity and maintained a high degree of uniformity. The advantages of developing otherwise worthless lines of inbred corn in order to utilize the heterosis consequent on their hybridization were readily apparent. And the same technique could be applied to other food crops.

In contrast to Burbank, Shull was very much a part of a continuous tradition. He followed in the footsteps of the Danish researcher Wilhelm Johannsen, who had studied inheritance in pure lines of beans, and both based themselves on the original discoveries of Mendel. Moreover, the full application of Shull's idea had to await the "double-cross" method invented by Donald Jones of the Connecticut Agricultural Experiment Station, who in 1917 successfully interpreted the phenomenon of heterosis in terms of the chromosome theory of heredity and proposed the use of four inbred strains instead of two. Between them, Shull and Jones accomplished one of the great feats of applied genetics, which was in the years to come to increase grain production by untold billions of bushels. The effects have been global and are scarcely calculable.

The period during which he observed Burbank was, Bentley Glass points out, the most fruitful of Shull's own scientific life. In addition to his work with corn, Shull was busy with studies of lychnis, digitalis, shepherd's purse, and the evening primrose. Between 1907 and 1910 he published forty-four scientific papers "in spite of his distractions in California." But he seems to have maintained a strict separation between his two worlds, Santa Rosa and Cold Spring Harbor. Around 1907, for example, Burbank was also experimenting with corn, crossing Indian corn (*Zea mays*) with the related wild grass teosinte (*Euchlaena mexicana*) to produce a hybrid that reportedly exhibited characteristics of both. Burbank deduced from this that teosinte was very likely the wild ancestral form of maize.* "It seems beyond understanding that the files of Shull's notes contain no comment of his own on this production," Glass complains.

> Shull's file on Zea [for his projected review of Burbank's varieties] is lamentably small and contains only a few newspaper and magazine clippings. These clippings do refer to the supposed maize-teosinte hybrid, but there is no comment from Shull upon the matter. Neither is there, anywhere in Burbank's materials, any indication that he knew of, or was interested in, what Shull was doing at Cold Spring Harbor. How amazing that there was really no meeting of minds in the very area in which it might most reasonably have been expected![25]

Shull and Burbank were both of them mild-mannered, puritanical in their habits, and capable of verbosity. "You say, 'He is always impatient of a conversation in which he does not do all or nearly all the talking,'" Woodward wrote to Shull. "Now, singularly enough, this is precisely the remark he has made to me concerning you."[26] And E. Carleton MacDowell noted that "throughout his life, Shull was never parsimonious in the use of words, either in speech or writing."[27] It is all the more peculiar, then, that Shull remained so closemouthed on the subject of Luther Burbank. Even with his immediate family he was uncommunicative and gave only the most concise answers to his children's questions about Burbank in later years.

Working with Burbank grew progressively more difficult for poor Shull. "Yesterday Mr. Harwood was here revising his book on 'New Creations,'" he wrote Woodward on February 26, 1907. "Dr. [William Mayo] Martin is here getting material for the ten volumes of the Cree Co., and a representative of Collier's was also securing data for articles, so you can see none of us will be able to receive much attention. Mr. Burbank is trying to treat us all alike and has assigned different hours to each of these various interests."

By January 1908, when the Carnegie Institution made its fourth payment of $10,000 to Burbank, Shull was beginning to wonder whether it was worth going on. Burbank's work was already being estimated at more nearly its correct value in the trade, he observed

* "It almost certainly derived from crosses between maize and Tripsacum, a native American grass, and therefore originated from maize rather than being its 'wild' ancestor as has been supposed," says Edgar Anderson in *Plants, Man and Life* (Berkeley and Los Angeles: University of California Press, 1971), p. 160.

in a letter to Woodward dated February 24. Doubtless the general public would in time also come to rate him more accurately even without any published report by the Carnegie Institution "as the work of other plant breeders becomes more widely known and appreciated." The tone of this implies an intent to debunk that was not part of his mandate. Shull was becoming irritated. He felt that the Cree people were receiving more of Burbank's time than he was and that one of the two projected studies was bound to be superfluous. Shortly after he resumed his task at Santa Rosa early in March 1908, he wrote Davenport, however, that, "I shall not find fault so long as he continues to take an interest in my side of the work, and shows the good spirit he now manifests." Burbank seemed in better health than usual, and Shull was optimistic of making good progress.

In August of that year, the institution sent Shull on a European tour of other plant-breeding establishments. He visited some thirty of these, from Copenhagen in the north to Naples in the south, from Dublin in the west to Brünn in the east. He gained an invaluable background in contemporary horticultural experimentation and also picked up current European opinions of Burbank.

English breeders seem to have been particularly critical. The Laxton Brothers of Bedford exemplified this attitude. Edward Laxton complained that "the exploitation of Mr. Burbank was wholly unfair to other plant breeders who had been doing excellent work for many years, and he thought that Mr. Burbank had accomplished practically nothing, as none of the plants that he had sent out have yet found their way into general culture in England or proved of any value there." This insular point of view disregarded the fact that Burbank had been working with varieties intended to do well in California's Mediterranean climate rather than in that of England. But the simple lack of charity to a fellow breeder was probably attributable to the hostility already being exuded by the American horticultural establishment.

Meanwhile, that same August, Woodward paid Burbank a visit to see what was going on. He found him "pretty badly tangled up with the universe, so badly, in fact, that it was necessary to issue a sort of ultimatum to him," Woodward wrote Davenport. "I hope we may be able to extricate him, in part at least, since I feel that he is making good to the Institution on the horticultural side. He has certainly achieved great success with his edible cactus." Unfortunately, there is no clue as to what the precise nature of the entanglement was. The ultimatum was a strong hint that the Carnegie grant might be discontinued if he did not straighten out.

Woodward remained personally sympathetic, and evidently Burbank also retained the support of Andrew Carnegie. At a meeting of the board of trustees on December 8, 1908, Carnegie observed that "if we can sustain Mr. Burbank in his work so that in the end we will have economic gain, I would even be in favor of increasing the amount given to him." (By "economic gain," practical benefit to mankind as opposed to purely scientific outcome is to be understood here, not profit on the part of the institution.) Woodward agreed that the scientific results were "purely secondary" and that the "economic aspects of the case" were almost "the sole justification for the undertaking." He noted, however, that he was often asked "why we, as an institution, are subsidizing a faker." In the upshot the board of directors went on record to the effect that it was

> desirous of seeing the work brought to an end. G. H. Shull is to return to Luther Burbank as soon as convenient and shall finish the work with Burbank by the end of 1909. Voted that the year 1910 should be devoted by Dr. Shull to writing up his investigations and gathering illustrations for his final report.

Writing to Burbank on December 21, 1908, Woodward informed him that a further appropriation of $10,000 for the year 1909 had been made but warned that "by reason chiefly of the numerous misrepresentations of your work in the popular press," he was having difficulty in convincing the board of trustees to continue its support. He expressed the hope that "in future we may be able to divest this work of its sensational features to such an extent that its real merits will plainly appear." The handwriting was on the wall.

13

THE LUTHER BURBANK PRESS

In January 1909, at the annual meeting of the American Breeders' Association, held that year in Columbia, Missouri, Burbank presented a paper entitled "Another Mode of Species Forming."* In it he reported a phenomenon science would be twenty more years in recognizing: the true-breeding species hybrid. Unfortunately, nobody believed him, and to this day his priority in reporting such hybrids has not been conceded. "Possibly this paper was ghost-written by G. H. Shull. It bears his fingerprints," the geneticist W. E. Castle, who seems to have been relying entirely on memory and got both its title and its contents wrong, later remarked unkindly.[1] Quite apart from the fact that Shull had been in Europe for the past four months, he would, had he accepted Burbank's findings on this score, almost certainly have made his own report. He did not, and the existence of what are now known as amphidiploids, or allotetraploids, species hybrids whose somatic cells contain the diploid chromosomes of both parents, would have to wait until the cytological studies of the 1920s identified them.

Meanwhile, as far as the rumblings from the Carnegie Institution were concerned, Burbank seems to have made few, if any, con-

* See Appendix I.

cessions. It was not in his character to do so, and as sales remained good, he was not overly concerned about money. In October 1908, the Thornless Cactus Farming Company had bought ten more varieties of opuntia from him for $5,460.50. Given this further endorsement of his personal faith in the eventual economic importance of his cacti, it would hardly have seemed necessary for him to go begging. A fifth—and last—Carnegie payment was made in January 1909. When Shull turned up in the spring to resume his work, there was no apparent acrimony. But another visitor that April, the popular philosopher Elbert Hubbard, observed that his "blue eyes . . . would be weary and sad were it not for the smiling mouth."

In February 1909 Burbank had had a new proposition, this one from the brothers Herbert and Hartland Law, owners of the Fairmont Hotel in San Francisco. The Law brothers, who had made their fortunes as promoters of patent medicines, seem to have been brought together with Burbank by the flamboyant Oscar Binner, a partner in the Cree Publishing Company who, though himself a professional promoter, was a true Burbankian and remained associated with his enterprises for a number of years. Burbank was persuaded to sign a contract, the terms of which established a corporation, headed by the Laws, to be known as Luther Burbank Products, Inc., which would market everything he produced. Walter B. Clarke was taken on as sales manager at Santa Rosa, but while the project was still largely in the planning stage, Burbank seems to have had second thoughts. He repudiated the preliminary contract, and fortunately the other parties were prepared to bow out gracefully if that was what he wanted.

Binner declared that "for myself and my associates, the Law Brothers, let me say that Mr. Burbank's absolute happiness and contentment are our first consideration." Clarke was placed in a somewhat difficult position, as he had given up another job to take up his post in Santa Rosa, and Burbank was induced to let him stay on. They do not seem to have hit it off too well. Clarke thought Burbank "erratic in his actions and slipshod in his methods . . . a strange combination of childlike simplicity and Yankee shrewdness," while Burbank for his part decided that he did not need a sales manager. Clarke's contract, too, was therefore terminated, with some suitable compensatory payment. This was the end of the first scheme to market Burbank in a big way as a purely commercial proposition.

In 1909 Governor James N. Gillett signed the bill making Burbank's birthday, March 7, State Bird and Arbor Day, as proposed by the California Federation of Women's Clubs three years before. But even as this honor was being paid him, a fresh storm was brewing.

In March 1908, he had sold John Lewis Childs a hybrid berry he called the "Sunberry." Childs had renamed it the "Wonderberry," with an eye to the market. According to Howard, it was an F_2 hybrid between *Solanum guineense*, the staminate parent, and *Solanum villosum*, the pistillate parent. Both progenitors produced fruit that was inedible, though not actually poisonous, but the fruit of the hybrid proved good, either cooked or eaten raw when thoroughly ripened. Having bought this berry for a mere $300 (or $375—there is some discrepancy in these figures), Childs proceeded to advertise it widely as a new Burbank marvel, going so far as to issue a recipe book entitled *100 Ways of Using the Fruit of the Sunberry or Improved Wonderberry*.

The controversy that followed has been described in detail by Charles B. Heiser, Jr., in the chapter on "The Wonderberry" in his book *Nightshades*.[2] Early in 1909, battle was opened in the pages of the *Rural New Yorker* (described as "the business farmers' paper"). The pet cause of this weekly journal's crusading editor, Herbert W. Collingwood, was honesty in advertising. He had been suspicious of Burbank from the start, Collingwood let it be known. But there was evidently bad blood here going back some years. In a letter to Vernon Kellogg at Stanford dated January 17, 1906, Burbank had noted that "Dr. Van Fleet of the *Rural New Yorker* . . . tried by every possible means to get me to take him into partnership, and got very angry because I would not do it."

It seems to have been in an article in the *Rural New Yorker* that the charge was first made that the Wonderberry was nothing but *Solanum nigrum*: the black, or garden, nightshade, a common weed throughout the United States (but not to be confused, however, with the deadly nightshade, *Atropa belladonna*). This accusation was reiterated on May 29, 1909, in an attack entitled "The Wonderberry and the Wizard Burbank," which jeered:

> I'm a very humble reader, with a home among the hills, and the thought of eating nightshade all my soul with anguish fills. I'm the ultimate consumer, and I only want to know, if the berry is a wonder, whether Burbank made it so!
>
> The "Wonderberry" appeared this season as one of the "novelties" which are sprung upon the public without official test or preparation. We had no chance to test it, but botanists of high reputation were sure it was in no wise different from the well-known *Solanum nigrum*.

The writer went on to cite a letter from Burbank offering $10,000 in cash to "any living person on earth" who could prove that the Wonderberry was the black nightshade. Since it was by Burbank's own admission a cross between *S. villosum* and *S. guineense*, and these had been declared to be "nothing other than forms of *S. nigrum*, a weed in every country," Burbank should "at once hand himself that $10,000 for he has earned it."

In July 1909, the credibility of the Wonderberry suffered a further blow when the judges at the Boston Flower Show declared it to be worthless. "Luther Burbank received the first severe snubbing yesterday when his latest creation the Wonderberry, or Sunberry, was declared a failure," said the Boston *Post* on July 18, 1909, compounding the confusion by identifying the berry as a hybrid produced by crossing *S. nigrum* with *S. africanum*. The *Rural New Yorker* kept up the attack. The Wonderberry was "mentioned or discussed at length in no fewer than 34 issues," Heiser observes. Collingwood's constant theme was "Do we earn that $10,000?"

Though Burbank was willing to bet that the Wonderberry was not nightshade, he had no easy way of establishing its true identity. His chaotic system of record keeping provided little assistance, and his lack of precautions in hybridizing meant that there was always the possibility that his experimental specimens had been fertilized by nightshade pollen from wild plants growing in the vicinity. It was likely that this had happened to Childs's nursery-lot multiplications, the common reaction to which was negative. As one Mississippi lawyer wrote Collingwood:

> Everybody in this section that bit at the fraud, has had the same experience with it—they got nothing but a black nightshade utterly unfit for man or beast to eat, altho they carefully matured it, and not only tried to use the "berry" when it first turned purple, but waited until it was thoroughly ripe, and never were they fit for a duck to eat, either raw or cooked.*

He offered the editor "the thanks of a grateful people" for exposing "Burbank's and Childs' fraud in this matter."[3]

Collingwood, in the upshot, thought that though "no one in the country ever expected Mr. Burbank would pay any $10,000 . . . it struck me at the time as it has struck most of the people who have heard about it as a very foolish proposition for a man like Mr. Burbank to make." A responsible horticulturist, in other words, ought not to throw the onus of botanical proof onto his customers when one of his products failed to live up to its advertising. "I don't know," the editor observed,

> of anything in recent years that has stirred up quite such a controversy as this "Wonderberry" proposition, but it is my judgement that Mr. Burbank has made a great mistake in letting the matter go as he has done. He has given the impression, to use a common expression, "That he bluffed without being able to make good."[4]

What was the Wonderberry? Heiser, a professor of botany and an expert on the nightshade family, made every effort to find out and concluded in the end that it was probably identical with a berry called *msoba*, which he obtained from South Africa. He tried some in a pie. "While they don't quite measure up to Childs' claims," he observes, "they are certainly not as bad as the *Rural New Yorker* would have had us believe."[5]

In August 1909, Burbank suffered another assault. A resolution passed by a meeting of the Pasadena Gardeners' Association condemned "the nature-faking methods and exploitations of alleged but false creations by Luther Burbank," deploring the fact that "a

* Though the unripe berries of the nightshade are definitely poisonous, the ripe berries are considered safe to eat when cooked. The nightshade family, or Solanaceae, to which the Wonderberry belongs, includes the potato, *Solanum tuberosum*, and the tomato, both of which contain toxic solanum alkaloids (solanine) in their green parts and were once rejected as food partly because of their botanical associations.

false impression has been given the public concerning plant breeding by Burbank." Defending himself in an interview with the San Francisco *Chronicle* shortly after this condemnation, Burbank protested:

> The extravagant estimates of my work have been the bane of my existence. There has been much written about me by sensational writers who know nothing either of me or my work. I am not responsible for all these things and anyone with any knowledge of horticulture could discern at once that much of the stuff sent out is nothing but the space-writer's chaff.[6]

But there was no getting away from the fact that he had encouraged those sensational writers; and he would continue to encourage them. For what, after all, were the grumblers in Southern California to him? That year he was complimented by the publication of *The Scientific Aspects of Luther Burbank's Work*, a book by David Starr Jordan and Vernon Kellogg that took him very seriously indeed, spoke of his genius, his ability to do things that few other men were capable of, and his extraordinary keenness of perception.[7] The authors were both prominent as scientists (Kellogg, then in his early forties, would later become permanent secretary of the National Research Council), and Burbank must have felt vindicated despite the carping in the trade. At home in Santa Rosa, he was elected park commissioner, found it necessary to open an information bureau to deal with the ever-increasing volume of inquiries from the public, and allowed the local ladies to persuade him to ride in the annual rose carnival parade.

And what of Shull? On August 26, 1909, "a very hot day," he married his second wife, Mary Julia Nicholl, in Brooklyn, New York. They had met on shipboard, returning from Europe where she had been traveling with her mother. But duty soon called. By November 1909, he was back in Santa Rosa, writing Davenport:

> Mr. Burbank has practically recovered from his long illness and has been able to finish up to the letter Z, but did not find time to go back over the earlier letters as we had hoped to do. He has asked me to leave with him the records I have of those earlier conferences and he will try to find some time to improve them this winter.

That winter, however, in mid-December, Burbank's mother died at the age of ninety-six. Her death could hardly have been unexpected, but they had shared the same roof for almost sixty years, and few, if any, relationships had been more important to him. Scarcely had he recovered from this blow when he received news that the Carnegie Institution was canceling his grant. In a letter dated December 27, 1909, President Woodward wrote:

> I regret to inform you that the Board of Trustees of the Institution, at their annual meeting, determined to discontinue subsidies in aid of your horticultural work.
>
> It is unnecessary here to set forth the reasons which have led to this action. The possibilities of such a determination were contemplated in drawing up the letter sent you under date of December 29, 1904. . . . The probability of such action was also indicated to you in the summer of 1908 on the occasion of my last visit to you.

Woodward concluded by hoping "that we may be able to complete for publication at no distant date the record of your work drawn up by Dr. Shull."

When Shull returned in May 1910, he found Burbank falling back on his earlier plea: he had no time to give him more help. "He says he has no income now, and that his time is worth $500–$600 an hour," he reported. Though Burbank was friendly enough to him personally and showed him his newer cultures and results (among them a Chinese climbing plum called *mao-li-dzi*,* which Burbank thought might be a new species), Shull decided that he could expect little or nothing more from him in the preparation of his report.

He now submitted the first completed section of this—dealing with the subject of rhubarb—to President Woodward, with copies to his colleagues Davenport and MacDougal for their criticisms. Since this was the only section he ever did complete, Shull's analysis is perhaps worth looking at in some detail.

Rhubarb seems to have originated in central Asia and was first cultivated there for medicinal purposes—the roots being used as a

* Presumably *Actinidia chinensis*, the Chinese gooseberry, or kiwi fruit; the name *mao-li-dzi* evidently means "hairy plum." The kiwi fruit was introduced to New Zealand about 1904, and commercial plantings in that country, the United States, and elsewhere are believed to derive from these first plantings outside the Far East. Burbank's own *mao-li-dzi* does not seem to have caught hold.

drug. By the latter part of the sixteenth century, the plant itself had been introduced to Italy, and it is known to have been grown near Padua. It reached England early in the seventeenth century, and as one of the earliest of fresh spring vegetables, and the only one that is a dietary substitute for fruit, it came to be prized in private kitchen gardens. During the nineteenth and early twentieth centuries, Shull notes, there was a considerable demand for it, in supplying which a number of fortunes were made. Later, with the increasing availability of fresh fruit, the demand slackened off, and today it is comparatively little eaten, though a good source of vitamins A and C, as well as of iron and calcium.

Like all economic plants in long cultivation, rhubarb had given rise to numerous varieties, and when Burbank began to study it, he found that the seedlings differed greatly in earliness, in thickness, length, and color of the stalks, compactness of growth, and cooking quality. He crossed a number of species but found that the hybrids did not compare with the best varieties already in existence, all of which apparently derived from *Rheum rhaponticum*, and perhaps also from *Rheum undulatum*, two of the species he was experimenting with.

Though it could be forced in greenhouses and cellars, the normal rhubarb season was then limited to a period of from four to six weeks in early spring. About 1890 Burbank heard of a new variety introduced in Australia and New Zealand called Topp's Winter Rhubarb, which produced stalks during the winter as well as in spring. Practically everbearing, it had apparently originated from a seedling raised by a market gardener named Topp at Buninyong, Victoria. It had a better flavor and a finer crimson color than the old varieties, but the stalks were small, and it quickly ran to seed.

Burbank immediately saw the importance of obtaining a winter-bearing variety and ordered some of the plants. The first two or three consignments failed, as they died aboard ship, but eventually, about 1892, he obtained about half a dozen small roots from Auckland, New Zealand. He multiplied this stock, and the seed was carefully harvested and planted over several years. All the plants raised reproduced true to their habit of winter bearing, though differing widely in other respects. The best individuals were selected each season, and by November 1, 1900, Burbank was ready to offer these as "a new vegetable—the Australian Crimson

Winter Rhubarb"—in a circular that, to emphasize its winter growth, had a photograph on the cover of a one-year-old plant of the new variety taken on Christmas Day. (Coincidentally, the same variety was imported into England from Australia about the same time, improved in much the same way by English seedsmen, and exhibited in London in 1903 as "Sutton's Crimson Rhubarb" and "Topp's Winter Rhubarb.")

In 1902 Burbank turned over his stock to John Lewis Childs for distribution, and Childs, as usual, lavished on it a too-glowing description—particularly with regard to its being "robust and perfectly hardy anywhere." Burbank was ultimately forced to correct this impression in a circular dated June 1909, declaring that "unfortunately I have not so far succeeded in producing a perpetual rhubarb which can profitably be grown in the cold Northern States. . . . *These new rhubarbs will not be profitable out of doors where the Eucalyptus, the Orange, and the Fig can not be grown.*" In Southern California, particularly, however, the Crimson Winter Rhubarb did wonderfully. One grower reported harvesting seven tons of stems from three-quarters of an acre set out the previous year. Results were so good that a grower at Whittier nicknamed it "the California mortgage-lifter."

Burbank went on with his work with rhubarbs and by seedling selection alone produced several other varieties. "Burbank's Giant Winter Rhubarb" was sold to Childs in 1905 for $1,500, and another variety, called "Burbank's Wonder Winter," was released in 1906. When Shull was writing, Burbank was still at work in a small way, "having at the present time several dozen varieties of large-growing plants which are in the process of being tested and which were offered for sale as 'Select Improved Giant Varieties Mixed' in a circular issued in June, 1909."

It is a success story, if not a very spectacular one, and is enhanced by the fact that this was only one of numerous varieties he was improving at the time. This must always be borne in mind. While other breeders attacked the isolated salients that took their particular fancy, Burbank advanced over the broadest possible front. One thing that clearly emerges from the rhubarb report, however, is that no magical abilities were involved: it was what anyone might have done—what others did do almost simultaneously in England—given native wit and enterprise.

Toward the end of 1910, Shull wrote Davenport: "I am not complaining about the Burbank work. It has been exceedingly trying at times, but it has also been very pleasant in many ways and I am sure that the experience and knowledge gained will always be useful to me, as well as a source of satisfaction." Burbank had been kind to him from first to last, he felt, though rather free in his expression of disapproval of his treatment by the institution. As to the report, Shull hoped "to have it finished before next spring."

Meanwhile, work on the rival publication was also somewhat bogged down. The Cree Publishing Company had planned a multivolume account of the man and his work, to be written by Burbank himself, with editorial supervision by the Reverend Dr. William Mayo Martin and scientific advice from Dr. Leroy Abrams of Stanford. Abrams was in the field as early as 1907, and Martin set up shop in the old pre-earthquake Burbank cottage with a stenographer to tackle the editorial work.

Progress was slow, for Burbank was tugged two ways. On the one hand, he wanted very much to have the books; on the other he could not easily provide his editor with the necessary material. Records were wanting, for a start. But the main problem was that the projected work was to glorify him in exactly the way Harwood had done—or rather more so. And he had scruples. Besides, he knew very well that the people whose opinions he most valued were not impressed by that kind of thing.

Charisma is a double-edged instrument. He was trapped by his admirers. Reflected in their eyes was the image of himself that they wanted to see. But he knew, at the same time, that it was not a true image. All his training was for honesty: being true to oneself. Anything less made the whole enterprise meaningless. But he had this power—and it had little to do with horticulture—of making other people see him larger than life. And he had this weakness: he enjoyed their praise far too much.

Only he could really write the books, and if he wrote honestly, they must deflate the legend. He could not lie and tell the public he was a wizard. Neither, however, could he tell the strict truth and disabuse them. So he vacillated, as he had vacillated in telling Shull his story. Martin and Abrams struggled with the vast mass of what *seemed* like material: the catalogues, circulars, and brochures—

over seventy of them by 1910, the books of clippings, the thousands upon thousands of scribbled notes. Somehow, when looked at more closely, the clouds of evidence on which their ten volumes were to be based dissipated into thin air, leaving behind only the mute proofs on the ground—impossible now to decipher with any degree of accuracy—and the man himself, who would not be pinned down.

After a while, Martin and Abrams gave up. Their places were filled by Oscar Binner, now operating as a company, "Luther Burbank's Publishers," at Santa Rosa, and by Wickson, who still cared and believed that this story should be told.

As for Burbank—the Boss, as they called him—he was endlessly busy, especially with cactus. There were critics to be answered and the struggle to obtain for his opuntias the fair trial they had never had. He was increasingly bitter.

> Nearly two hundred and forty thousand dollars of my own private earnings in other lines have been used in this work, not one-tenth of which has ever been received in return, nor was it expected, as few originators of new plants have ever made it pay themselves no matter how many untold millions their work may have been of benefit to others,

he wrote in a pamphlet entitled *How to Judge Novelties*, issued in January 1911.

> No patent can be obtained on any improvement of plants, and for one I am glad that it is so. The reward is in the joy of having done good work, and the impotent envy and jealousy of those who know nothing of the labor and sacrifices necessary, and who are by nature and cultivation, kickers rather than lifters.
>
> Happening however to be endowed with a fair business capacity I have so far never been stranded as have most others who have attempted similar work, even on an almost infinitely smaller scale.

As an introducer of so many varieties of plum, Burbank was the star attraction at the Fruit Grower's Convention in December 1911. A large portrait of him, appropriately draped, hung over the president's chair. Burbank modestly took the last chair in the last row, flanked by George Roeding, the "father of Smyrna fig culture in California," and the faithful Oscar Binner. "When Mr. Burbank was finally called upon by the Chair and he stepped up to the platform," Binner noted, "I had my doubts as to his talk being a success, but, after the first question had been fired at him, and I saw with what ease he answered it and how he kept this up from Cactus to fodder, alfalfa to plums, from berries to apples and so on, it was a revelation to me."[8]

Binner was hard at work putting together and classifying the illustrations for the books now. He thought that the project was progressing well, but was anxious "that what is being done is done *right.*" In a letter dated January 21, 1912, Wickson warned him:

> You must always be looking out for "bearings." I will fix the spots you indicate and the author may pass it without your raising an issue. As you indicate he is really in conflict with himself and unconsciously with the history of plant breeding also. I think I can bring out his real meaning better than he does himself for I think I know what he means. He naturally does not wish people to get the idea that such plant breeding as he has done is easy and so uses some phrases to which the general reader would give too much technical significance. I am sure I can fix it. Keep on the line you are following. Shoot the thing all full of holes if you can!

Burbank, Binner reported back on February 5, was now "ripe" for pushing the books to completion. Usually it was the other way around, but now he was "itching" to hurry them along. Since he could not be pushed or driven, Binner observed, it was essential to cash in on this frame of mind. He asked Wickson to come up to Santa Rosa for a few days to go over the work in detail with Burbank. "He feels that he is in the dark—groping about in a mass of stuff, some of which may be good; some he feels would not do. He is uncertain and wants to know *definitely*, and I know that *you* can give him the desired information."

Evidently, however, the work was not being accomplished speedily enough to satisfy the Cree Publishing Company. Early in 1912, the project changed hands. A new firm, the Luther Burbank Press,* was organized by John Whitson and Robert John, with the latter as president. Cree bowed out of the picture, handsomely compensated for a company that was over $30,000 in the red.

Even before this, however, Whitson and John had set up the

* Incorporated under the laws of the state of Maine, May 23, 1912.

Luther Burbank Society (incorporated on April 3, 1912) with a view to promoting the forthcoming books. Membership in the society was to be limited to five hundred, with the first hundred to be regarded as "charter members." Invitations to prospective members were sent out, a $1 subscription being requested, and the publishers promised to send proofs of the forthcoming work for the members to criticize and help edit. The procedure was to be thoroughly democratic, but the ten volumes would each cost the subscriber $15. It was a not untypical scheme, bearing all the marks of the most unscrupulous type of exploitation. "My own invitation," says Howard, then a young instructor in horticulture at the University of Missouri,

> stressed the importance of quick acceptance as it was pointed out that only a few of the most important people of the United States were being invited and that I had the honor of being one of the number. To emphasize this point the invitations bore serial numbers. Mine was somewhere in the seventies. With the feeling that an obscure instructor had been mis-catalogued, I dropped the invitation in the waste basket. In a month or two a second invitation came along, this time bearing a much lower number—thirty-five or forty or thereabouts. This went the way of the first.

These invitations evidently went out to horticulturists and botanists at institutions throughout the country. If a plan had especially been devised to prejudice them against the very name of Burbank, it could scarcely have been more effective.

In terms of the original arrangement with Whitson and John, Burbank was to have received an advance of $30,000 and additional royalties on books sold. Apparently, however, he was paid only a small part of this, perhaps waiving the balance for the moment in view of the financial exigencies of the operation as a whole. Meanwhile, a bond issue of $300,000 to finance the work was advertised in the Santa Rosa *Press-Democrat*, commencing on November 3, 1912. The Luther Burbank Press, according to the full-page advertisement, "was originally set up by a group of prominent people from all over the United States who subscribed a total of $600,000 worth of stock—fully paid up, $480,000 preferred and $120,000 common." It was, they claimed,

> the biggest business undertaking of its kind not only in Santa Rosa but anywhere west of New York and will turn over several million dol-

> lars a year at a profit of nearly half the turnover. . . . The Company's securities have not been offered to the general public heretofore, and the present offer is confined to Sonoma County for special reasons, namely: the company will need the cooperation of Sonoma County residents in securing larger postal facilities from the Post Office Department, in petitioning the Interstate Commerce Commission for more equitable freight and express rates (since, when the delivery of the books commences, the company's freight and express bills will exceed $100,000 a year), in securing from the Department of Agriculture unpublished or private statistics or information, and from the Panama Pacific Exposition better than ordinary concessions.

The literature of the society contained explicit criticisms of the "other report"—which, it implied, had been dropped. "The Carnegie Institution of Washington appropriated a large sum of money for the promulgation of Mr. Burbank's discoveries," reads an invitation to life membership issued in 1912. "But after several years of effort, this project was abandoned, because it was the purpose of the Carnegie Institution to limit the field of writing to pure science, whereas Mr. Burbank's steadfast ambition was to give the benefits of his life struggle to the many instead of the few." (This by means of books priced at the then considerable amount of $150 a set.) "It is the *clear, practical* exposition of *everyday* methods which the Carnegie Institution would have made secondary to theories and science," the invitation adds in a further appeal to populist sentiment.

On September 1, 1913, a professional writer, Henry Smith Williams, was hired to produce the clear, practical exposition, replacing John Beaty, who himself had succeeded Martin, Abrams, Wickson, and Binner on the job. Williams was an experienced compiler of semiscientific books, including several encyclopedias. He now set to work to pull together the material accumulated by his predecessors (said to amount to some 1.5 million words) with all due speed. This he did so efficiently that by November 25 the first volume was at long last ready for the printers. Over $236,000 had apparently been spent on "educational propaganda, materials and publications to date."

Already the promoters were eagerly anticipating the expansion of their scheme. In a brochure directed at potential investors, they looked forward to five special editions in addition to the first. These were to be:

1. Popular edition in 8 volumes to be sold to 7,000,000 farmers and 4,000,000 rural and suburban people—50,000 sets at a net profit of $10 a set.
2. The text book edition, consisting of a series of 8 books, 2 for primary, 2 for grammar, 2 for high school, and 2 for college grades—17,000,000 copies annually, $1,200,000 at a profit of $200,000.
3. Special reinforced library edition, 8 volumes, same as popular edition except bound especially strong for frequent handling—15,000 copies. The Burbank books are a necessity to the libraries and their entire demand can be supplied within two years at a profit of $300,000.
4. The special Monograph Edition, consisting of a series of approximately 3,000 bound bulletins and booklets to sell at 10 to 50 cents, each dealing with one specific subject, to solve the soil tiller's specific problems and to give him precise information on specific subjects, the information to be collected from reports of the United States Department of Agriculture and of the various states. An estimated business of $250,000 a year can be developed at a profit of about $50,000 a year.
5. Foreign edition, of 8 volumes, same as American Popular Edition but translated into French, German, Italian, Spanish, Japanese, and Russian, and adapted for sale in England, Australia, South Africa, and Canada. Total market as great as for the United States.

It is hard to believe that anyone could have been taken in by this amazing farrago of rapaciousness and optimism. Howard has pointed out that edition number four, the so-called "Monograph Edition," amounted to nothing less than a proposal to cash in on the agricultural publications and bulletins issued *free* by federal and state governments, the presumption being that, with Burbank's name attached to them, people could be persuaded to pay for what they could otherwise obtain gratis from the Department of Agriculture and the experiment stations. But evidently many people were sufficiently impressed. The mail-order business was booming, so much so that the Santa Rosa post office had to be reclassified upward more than once to cope with the volume generated by the activities of the society.

As it happened, the Carnegie Institution had by no means abandoned the idea of producing its own report. Shull was busy on work of his own during most of 1912. In a letter to President

Woodward, dated August 20 of that year, he observed that he had always felt it better that "the Burbank book should be backed by a scientific reputation made by original research than that my research should be estimated on the basis of a reputation created by the Burbank book." In September 1913, however, he felt ready to resume writing up his findings. He had other reasons for wanting to spend some time in Germany, and on September 10 he sailed for Europe with the intention of settling down in Berlin to finish the job.

14

PIRATES AND PROMOTERS

Just how much Burbank himself had to do with the Luther Burbank Society and Press is a matter of conjecture. Presumably he was not at all involved in their promotional activities, which clearly bear the stamp of different hands. He doubtless gave Williams what help he could and perhaps dictated some new material. But the new editor was not afflicted with the hesitations of his predecessors. He had before him the material already assembled by Wickson, Binner, and Martin and his helpers, as well as some originally given to Shull. His instructions were to get on with the job and produce a readable—and salable—account. Where expansion became necessary, he found it easy enough simply to fill up the gaps, putting his own words into Burbank's mouth and occasionally even emerging into the text himself (it was supposedly written by Burbank) and referring to the subject in the third person.

That Burbank—who had always resented interlopers on his property and the interference of outsiders in general—was prepared to put up with this is evidence of his state of mind at the time. Williams left him alone. He was not always after him to supply answers to unresolved difficulties. For the moment, Burbank was satisfied with that. He badly wanted to be free of the secondary entanglements that took him away from his real work. Business—and that apparently included literary business—was no pleasure to him. It was this desire to be free that now led him into a second debacle, which in damage to his reputation outdid even the John-Whitson scheme. This was a new version of the project originally mooted by the Law brothers: a company that would assume responsibility for the marketing of all Burbank's products—past, present, and future—and free him to get on with the great, perhaps divinely appointed, work that it was his to do.

The Luther Burbank Company—no connection with the society and press that were being simultaneously promoted—was incorporated in California on April 22, 1912, by Rollo J. Hough and W. Garner Smith. (Smith, a young San Francisco stockbroker and insurance salesman, was brought in by Hough, an Oakland banker who was the main motivating force behind the scheme, to be secretary-treasurer of the organization.) The company agreed to pay Burbank $300,000 for exclusive rights to sell all his products, $30,000 in cash and the balance at $15,000 a year. It was provided that he should have the right to name the president of the company and that his attorney, through whom he would be able to approve the other members of the board, be made a director. The lawyer in question, F. S. Wythe of San Francisco, apparently found nothing wrong with the proposal. The contract was accordingly signed, James F. Edwards, a past mayor of Santa Rosa, was named president, and Burbank received his $30,000 down. Stock sales commenced in what appeared to be a very attractive investment. Ultimately, some $375,000 in stock was to be sold to the public at par, purchasers including a number of leading San Francisco bankers, merchants, and investment brokers.

The main trouble with the scheme was that neither Hough nor Smith had any experience in the highly specialized nursery and seed business. It seemed to them, however, that given the overwhelming public enthusiasm for Burbank's products and the great respect his name still commanded, they could hardly go wrong.

Now Burbank seldom had more than a comparatively small stock of his new varieties ready for marketing. It had been his practice previously to sell these to established nurserymen, such as Childs of New York, who would then propagate, advertise, and distribute the product. Here there were hazards, as the buyer might fail to maintain the integrity of the stock or indulge in unscrupulous advertising that brought the originator's name into disrepute, as had happened with the Wonderberry.

A corporation named the Universal Distributing Company was now established to take over this function. Nurseries and seed farms had to be set up to multiply the stock, if it were to be sold in large quantities, and this seems to have called for an expertise the company lacked, at least on the managerial level. To complicate matters, it started out with one of the most controversial of all Burbank's products, the spineless cactus, which was virtually the only thing offered for sale in 1913, the first year of actual operation. The first general catalogue, issued in 1914, advertised about one hundred varieties of seed and bulbs as Burbank productions and went on to list thirty pages of standard varieties. All but a few of the Burbank varieties had previously been offered in his own earlier catalogues. Seeds of the standard varieties were apparently purchased in bulk from an eastern firm. The fact that these were listed in a catalogue entitled *Burbank Seed Book* probably had the effect of persuading many buyers that they were the work of the master himself.

Sales policies were determined by a man named Hoffer, whose standing instructions were that all orders were to be filled *regardless of whether or not they had the proper stock on hand.* How this might be done was illustrated by Fred Suelberger, for many years horticultural commissioner of Alameda County, and in that capacity responsible for the inspection of the shipments of plants that passed through the company's warehouse in Oakland for disease and insect pests. Suelberger noted that

> great quantities of so-called spineless cactus [were] brought in from cactus farms in Livermore and other places to be shipped out, presumably for filling orders. The cactus leaves or "slabs" were tied up in bales and I particularly recall that Company employees were kept busy removing the spines with wire brushes, which seemed dishonest as the cactus was supposed to be free of spines.[1]

Needless to say, the fraud was detectable as soon as the plants began to grow, and the indignant buyers invariably gave Burbank credit for having perpetrated it.

The buyers were not the only ones who were disillusioned.

Early in 1914, Smith made a sales tour of the eastern and middle-western states in an attempt to drum up business for the company. On his return, most of the orders he had obtained were turned down, as there had been a decision to raise prices in his absence. Smith, at least, had started out with considerable faith in his product, so much so that he made a collection of all the Burbank novelties he could get and sent them to be planted on his father's farm in Kentucky. Everything was planted carefully, but the results were highly disappointing. This is indicative of the poor quality of the stock being multiplied on the company's farms and also perhaps of the fact that Burbank himself was no longer as reliable as he once had been. He was surrounded by charlatans at Santa Rosa, and his dealings were increasingly with men whose sole interest was to sell his name and reputation for whatever they could get. Meanwhile, his true friends—Wickson among them—were falling away in disillusion.

But his fame, increasingly souring to notoriety though it was, continued to spread. In 1912 he traveled to British Columbia with a party promoting the Panama-Pacific Exposition to be held in San Francisco in 1915. He was honored at every stop and was personally greeted at Victoria by the Canadian minister of agriculture, who had apparently come all the way from Ottawa specifically to do so. Earlier that year, on February 29, while the House of Representatives was considering the bill making appropriations for the Department of Agriculture for the fiscal year ending June 30, 1913, California Congressman Everis A. Hayes had introduced a bill authorizing the Department of the Interior to turn over twelve sections of desert or semidesert land—totaling 7,680 acres—to Burbank for further experiments with spineless cactus. Hayes noted that "in certain quarters there has lately been manifested a desire to belittle this great man and his work." He observed that "a man connected with the Department of Agriculture—I purposely forbear mentioning his name*—has seen fit recently to assail Mr. Burbank and to minimize, even to ridicule, his genius and the great work he has done and is still doing to increase the prosperity and happiness of his fellow men." Teddy Roosevelt was quoted for the

* This enemy was undoubtedly Dr. David Griffiths, senior horticulturist of the U.S. Bureau of Plant Industry, one of the chief critics of the spineless cactus promotion.

defense as saying: "Mr. Burbank is a man who does things that are of much benefit to mankind, and we should do all in our power to help him." Attention was drawn to the fact that "99% of the plums shipped out of California are of varieties originated by Burbank and practically all the potatoes." The bill passed in the House, only to be defeated in the Senate.

That Burbank himself did not lose his sense of humor over the rancorous issue of his cacti was testified to by Cecil Houdyshell, who spent some days with him in the spring of 1913 and noted: "I hung around all day. Personally I get rid of people who stick around like I did. But Burbank was always affable, patient and very friendly. Never the unapproachable 'big' man." One afternoon after lunch, while Burbank was taking his nap, Houdyshell roamed around the gardens at Santa Rosa eating spineless cactus fruits, which he liked very much. When Burbank came over from his new house across the street, Houdyshell was busy trying to remove the tiny spicules that had stuck in his hands. "Ah ha!" Burbank exclaimed. "I see you've been into my spineless cactus." And he laughed at his joke, as much at his own expense as at his visitor's.

Shull, meanwhile, was in Berlin, his energies variously absorbed by his own work and by the task of writing up his Burbank findings. The Carnegie Institution had subscribed to the Williams book on his behalf, though Davenport noted that "the whole thing impresses me as a fake, particularly the way they write those personal letters which are identical with the printed forms."

In February 1914, proofs of volume 1 arrived in Berlin, forwarded from Cold Spring Harbor. Whether this was by special provision (Burbank being particularly interested to know Shull's opinion) or normal procedure is hard to tell. Other subscribers do not seem to have received proofs or been allowed to participate in the editing of the text, as promised in the prospectus. Judging by the proofs, Shull wrote Davenport:

> It appears to me a criminal waste of good paper, and the "science" involved in the text is certainly amusing. But it is not for me to criticize; it remains to be seen whether I can do any better. . . . The colored plates will prove both interesting and valuable. Aside from the financial considerations, it is chiefly for these plates that the book will exist.

When, in June 1914, the first three completed volumes arrived, with the immediate discovery that "considerable sections are almost word for word the same as my manuscript," Shull wondered whether it might not be better for him simply to abandon his own report. The Carnegie Institution might rest content "with having supplied the incentive as well as collected the data for so good an account of Mr. Burbank's work as this."

If Shull was understandably bitter, there are also clear indications that he was having trouble with his own writing and would have been happy to use this pretext as an excuse for dropping the whole thing: "As you know," he wrote Davenport:

> Dr. Williams is a much more gifted writer than I am, so that it seems to me wasteful to spend more time on a work whose chief difference lies in a less elegant style of presentation and more meagre illustrations. I am quite convinced that most of the value of the work has been attained through the competition with the Carnegie Institution, and I am personally content with that result if the Institution should decide to go no further with its report on Mr. Burbank's work—only, I think that in that case the sooner we decide to drop the matter the better.

Davenport, on the point of leaving for Australia, cabled a succinct reply: "Stop." On June 28, the day before Shull's letter, the Austrian Archduke Franz Ferdinand and his wife had been assassinated at Sarajevo. World War I was about to begin and would be offered in the Carnegie Institution's annual report for 1914 as the reason for not finishing the report.

But the excuse was a thin one. Even if all the facts included in Shull's notes were contained in the Williams books, they were so embedded in boastful adulation as to be useless. As Burbank had told Shull early in 1907, the ten volumes—now expanded to twelve—would not contain as much "meat" as ten pages of the Carnegie report. Shull had not merely accepted Burbank's dictation. He had studied his work objectively with the trained eye of a geneticist. He was, perhaps, the only fully qualified person to have done so in detail, and had devoted perhaps half his time in the preceding eight years to this report. Now he proposed simply to abandon it on the pretext of the European conflict, which had small conceivable relevance.

He drafted a sarcastic letter to Burbank from Berlin on August

1, the day Germany declared war on Russia, in reply to one giving some information he had requested on plum names.

> I was very much pleased to hear that your new books are being so well received, and congratulate you on that. I too have read them with a great deal of interest of course, and appreciate very much the charming manner in which you illustrate the Mendelian principle. I have noted several slight oversights, which you will be glad perhaps to call to the attention of your publishers in order that they may be corrected in subsequent editions . . .

Here the draft breaks off. Was it ever sent? What were the corrections? At this late date, it is unlikely that we shall ever know.

In 1941 Shull elucidated the position, writing in response to Howard's request for access to his Burbank notes. "It was a decision of the Carnegie Institution, not myself, that no report should be made until after Mr. Burbank's demise, but I was in entire accord with President Woodward on this point," he wrote.

> With the passing of time the urgency for the publication of a report on my experiences has subsided and it has not seemed wise to divert my time from my own fundamental researches to reanalyze the material and decide how much of it is worthy of publication. This reanalysis can be satisfactorily done only by myself and must wait until the time when my time and energy are not more advantageously devoted to my own special research problems.

Shull still felt that he had "a large personal equity" in the accumulated material resulting from his years with Burbank and that he would rather make the final report himself than have it made at second hand by Howard. The latter was, therefore, denied the opportunity of making use of this unrivaled source of information in preparing his definitive monograph on *Luther Burbank's Plant Contributions* and was also prevented from clarifying the role of the Carnegie Institution, as he would have been able to do had he seen the Shull-Davenport-Woodward correspondence and the minutes of the Carnegie Committee on Burbank.

There remains another possibility, hinted at in Williams's volume 12. This is that Burbank himself had the Shull report suppressed. How he might have done so is a matter of pure conjecture. He did have influential friends, and some sections of the Shull notes that he may have seen could well have been extremely displeasing to him. "Most great undertakings experience a number of false starts before they are finally launched on their way to accomplishment. And such was the case wth the beginnings of this work," Williams observes portentously.

> Laboring alone, and many years in advance of his time, it was not to be expected that Luther Burbank could be interpreted in the language of contemporary science. And in fact, with true Yankee keenness, he much preferred that his benefits be reaped directly by those who practice agriculture, rather than by those who merely study agriculture. If both classes could be equally benefited, well and good; but if one had to be slighted, then let it be the studier and not the practicer.
>
> With a willing heart, the able men appointed by the Carnegie Institution co-operated with Mr. Burbank, and the magnitude of the task, whatever the viewpoint, became apparent as page after page of manuscript was boiled together into what promised to become an interminable record.
>
> After a number of years Mr. Burbank saw and keenly realized that the work which had been done fell far short of his ideals—whatever its scientific value, it failed utterly to be the crystal clear presentation for the benefit of the practical man, which had always been his guiding ideal. So the first step toward success ended in what, at the moment, appeared to be but an expensive failure.[2]

George Shull died at Princeton, where he was professor emeritus of botany and genetics, on September 28, 1954. His work in the development of the pure line method of corn breeding had resulted in increased yields of from 25 to 50 percent, adding more than half a billion dollars a year to the value of the U.S. corn crop without increasing acreage. Shull received no material recompense for his discoveries. He was content, he said, with the knowledge that he had made an outstanding contribution to the welfare of mankind: a Burbankian sentiment if ever there were one. Though his home had been at Princeton for almost forty years, he chose to be buried at Santa Rosa. His wife, Mary, was buried there beside him on her death twelve years later.

The report on Burbank's work, begun so long ago, was never completed. Probably, as Shull observed, the reanalysis of his notes could have been done satisfactorily only by himself. When these notes were studied by a colleague, Dr. E. Carlton MacDowell, in the late 1950s, the latter concluded that "the real reason that one of the first major projects of the Institution bore almost no fruit, can only be suggested."

It may be that it was a comparatively simple one: Shull, with all his diligence and expertise, found it an impossible task. It is true that he had his own work to do, and by 1914 he was already aware of how important this was likely to be. In later years he suggested that this was the overriding reason, saying that he had

> determined from the start and with the full understanding and consent of the administration of the Carnegie Institution, that there would be delay in publication of the results, certainly until after Mr. Burbank's death. My own insistence on delayed publication was for the somewhat selfish reason that I was unwilling to have my reputation as a scientist rest to so large extent as it would have done if early publication had been required, on a work which even under the best of circumstances could not have had the high scientific value of a piece of personal research of my own.[3]

But this smacks of a saving afterthought. During the time that he was at Santa Rosa there is no hint of a suggestion that there was to be a deliberate holding back of the report until after Burbank's death. It seems far more likely that he simply could not write it. Even he could not unscramble the true from the false, the achievements from the exaggerations.

The twelve weighty volumes of *Luther Burbank: His Methods and Discoveries and Their Practical Application*, "prepared from his original field notes covering more than 100,000 experiments made during forty years devoted to plant improvement, with the assistance of the Luther Burbank Society and its entire membership, under the editorial direction of John Whitson and Robert John, and Henry Smith Williams, M.D., LL.D.," began to appear in 1914. Published by the "Luther Burbank Press, New York and London" and copiously illustrated with "direct color photograph prints produced by a new process devised and perfected for use in these volumes," they were undeniably handsome, with ornate leather bindings, deckle edges, and heavy paper watermarked "Luther Burbank Press" in large Old English. The hundreds of color photographs were remarkable for their time, and if some were of such inconsequential subjects as the boxes used by Burbank as seed flats, others were frequently skillfully chosen and composed. If editor–ghost-writer Williams lapsed into the absurdity of attempting to prove that rhubarb introduced from the Southern Hemisphere henceforth followed the calendar rather than the seasons and vainly contended that Burbank had hit on Mendel's laws before the rediscovery, this probably went unnoticed, save by a handful of horticulturists. Elaborate descriptions of ordinary techniques of grafting and budding, which had nothing particularly Burbankian about them, were included, and doubtless made an impression on the lay reader, who might suppose that these were part of Burbank's contribution to horticultural science. Williams carried the technique of padding—more widely practiced in his day than in our own—to a high art. Even now, one cannot fail to be impressed by his skill. But that, of course, was precisely what he had been hired for.

In January 1915, it was announced that the twelve-volume set was complete and that the headquarters of the press was being transferred to New York—despite earlier assurances that "the Santa Rosa office is, and will continue to be, the principal place of business." In addition to the elaborate subscribers edition, a somewhat more functional one was evidently brought out for library and institutional use, also in twelve volumes, and plans were underway for the promised popular edition, condensed into eight volumes.

The bubble burst abruptly. Perhaps demand (there seems to be no surviving record of how many sets were published and how many sold) did not meet expectations. At least $400,000 had been spent on the project, and the press was evidently financially overextended. The whole enterprise seems to have collapsed in a matter of months. The announced move to New York was probably a preliminary precaution on the part of Whitson and John, who must have known what was coming. Those who had bought stock lost their investments, and needless to say much of the blame was laid at Burbank's door, though he himself claimed that the company owed him $100,000 when it closed down. On March 12, 1916, the press forfeited its charter to do business in California by reason of nonpayment of taxes, and the promoters (if they had not already done so) went back into the woodwork.

Burbank now had his books—he retained copyright—but closely examined they were a strange hotchpotch. It seemed obvious to Shull that his own notes had gone into the compilation of the text, sometimes almost verbatim, and the community of horticultural and agricultural scientists—of which Shull was an increasingly prominent member—probably knew it too before very long. Shull had noted, for example, that on April 20, 1908, Burbank had told him:

> Now I have what I believe to be the best collection of Amaryllis on earth. At least all those that know the Amaryllis in Europe and America who have visited my grounds pronounce them far superior to any they have ever seen, *not only in size, but rapid multiplication and general effectiveness.*

And page 98 of Williams's volume 9 had this to say:

> From a mere horticultural standpoint, it is considered by experts to be the best collection of Amaryllis in the world. Not only has this colony the greatest diversity of forms but the most extraordinary individual plants.
>
> Experts of both Europe and America who have visited my grounds are agreed in pronouncing my galaxies of Amaryllis far superior to any to be seen elsewhere, *not only in size but in rapid multiplication and general effectiveness.* [Emphasis added.]

From Burbank's personal point of view, one of the most grievous casualties was his friendship with Edward Wickson, who had put a good deal of work into preparing the earlier draft of the books and received no acknowledgement. He, too, now joined the ranks of the disillusioned. That Burbank was painfully aware of this is testified to by a letter in protest dated April 20, 1916. Formally commencing "Dear Sir" instead of the old "Dear Friend Wickson," Burbank observes:

> I used to feel that I had your sympathy and confidence, but of late I know that I have not—not from what you have written, but from the attitude you have taken. Frankly—I had given you credit for having discrimination enough to know that the treatment which you received from the pirate book publishers was not Luther Burbank and I judge that their treatment of *you* has changed your attitude towards *me*. Future events will prove or disprove this.
>
> I have no possible feeling towards you otherwise than the kindest

> and have been surprised and hurt at the attitude which you have taken on several occasions during the last year or two.

But it seems to have been too late to mend this fence.

Meanwhile the Luther Burbank Company was also moving toward catastrophe. Early in 1914, dividends as high as 12 percent had been paid to stockholders. Heavy sales of cactus provided momentum, but these inevitably fell off when the company's sharp practices began to come to light. Financial difficulties commenced, and that spring saw the resignations of James Edwards and Garner Smith, both of them presumably disgusted by sales policy. They did not relinquish their stock, however, as they would have been well advised to do, and are said to have ultimately lost some $100,000 between them when the crash came. Dividends soon dropped to 6 percent and then ceased entirely, and Burbank himself was obliged to accept promissory notes instead of cash.

Finally, in December 1915, Burbank sued to recover $9,775 owed him. He had long delayed action—some of the stockholders were, after all, close friends. "Burbank has been the victim of stock pirates," one of his attorneys now declared. "This Company was formed three years ago. He took no stock in it and no interest in it. Some of the best men in town have also been made victims." The contract was declared void, further use of Burbank's name was forbidden, and another suit for $15,000, also in arrearages, was threatened. On February 8, 1916, the company was declared bankrupt, stockholders receiving little or no compensation from remaining assets, and in July of that year Burbank gave notice in a leaflet that he was resuming full control of his nursery and seed business himself.

Pirate book publishers . . . stock pirates . . . he had been exploited all round. One of his secretaries, Pauline W. Olson, a local woman who worked for him on and off between 1905 and 1912 and subsequently, thought him to have been a poor businessman, "careless, sloppy and unsystematic in his business affairs." Despite Burbank's own assertions to the contrary ("happening to be endowed with a fair business capacity . . ."), she was probably right. He was shrewd enough but lacked sufficient interest to manage business affairs on this scale well.

But though stockholders and scientists might be disillusioned, there is no indication that the general public had fallen out of love with him. And 1915 provided him with a propaganda countercoup to match his defeats. That October, almost at the peak of his troubles, Henry Ford and Thomas Alva Edison, visiting the great Panama-Pacific Exposition in San Francisco, took time out to pay their respects, traveling to Santa Rosa in Ford's private railway car in order to do so and bringing in their entourage a small party of notables, among them Harvey Firestone.

It was no impromptu call. Burbank had extended the invitation to them in San Francisco a few days before, and arrangements had been made to receive them in proper style. The local paper announced the event in banner headlines, and the newsreel cameramen had been forewarned and were there to report it. (Burbank seemed to enjoy the new medium, readily "performing" for the "moving photo operators" among his spineless cactus, film that had wide circulation on the Pathé news.) Massed school children—in those days few opportunities were lost of dragging them into the act—had been assembled to pay homage to the three famous men. Everything had been stage-managed to perfection.

15

"I LOVE EVERYTHING!"

Since late 1914 Burbank had had a new private secretary, Elizabeth Jane Waters, who was eventually to be his second wife. Though her people had originally come from Massachusetts, and there was some talk of her being distantly related to the Burbanks, she was a Michigan girl with a strict Seventh Day Adventist background. On coming to California, she first went to San Jose, where she had a sister. Later she moved to Santa Rosa, bringing with her a child, Betty Waters, who was a few years old at the time. Elizabeth started work with the Luther Burbank Press but presently transferred to Burbank's own office, displacing Pauline Olson, who had worked for him for more than seven years. Now she was there to share the duties of entertaining Ford and Edison with Emma Beeson—who can scarcely have appreciated the division of honors.

Elizabeth Waters was henceforth increasingly to be found at Burbank's side, taking a prominent place again in April 1916 when he was a guest of honor at a local parade (he had evidently not entirely sickened of parades). It was probably no surprise when they were married on December 21, 1916, by the pastor of the First Unitarian Church of San Francisco. But, as such marriages do, it certainly set tongues wagging; she was then in her mid-twenties, he was sixty-seven. Their honeymoon, the first real holiday Burbank had had in a long time, was a week in San Diego at the home of a friend.

Meanwhile, despite all criticisms, the publicity mechanism rolled on: once started it could perhaps no longer be stopped. In 1917 the California State Board of Education published a booklet on *Conservation, Bird and Arbor Day* that might have been deliberately designed to idealize Burbank. It came into wide use in elementary schools all over the country, with predictable effects on the generation then attending them. Even among his critics, there were many who did not want to see Burbank debunked. George Shull succinctly voiced this opinion:

> It is my impression . . . that he was a man of the finest, cleanest character of any person I have ever known. I always felt that he was the sort of man who *deserved* to be a popular hero. . . . Because of his clean life and formally expressed high ideals I have always wished to avoid saying anything which would tend to take Mr. Burbank off of his popular pedestal. The ideals expressed in his various writings on childhood, and the advertised motivation of his life work have a highly inspirational significance.[1]

He was above all a hero of his time; and only time itself would dethrone him.

In April 1917 the United States entered the war against Germany and her allies, and although Burbank was an ardent pacifist, his reaction was much that of his friend David Starr Jordan, who told the San Francisco *Bulletin*:

> *Our country is now at war and the only way out is forward.* I would not change one word I have spoken against war. But that is no longer the issue. We must now stand together in the hope that our entrance into Europe may in some way advance the cause of Democracy and hasten the coming of lasting peace.

In 1918 Burbank was to be found listed as a member of a "National Emergency Food Garden Commission," which aimed at inspiring the planting of a million home food gardens. On the business side, he introduced several new varieties of grain, promoted as the answer to the food shortage provoked by the European conflict.

One of these was a wheat he called "Quality," seed of which was sold for $5 a pound, or $300 a bushel. After it had been grown for several years, it was found to be identical with an Australian wheat called "Florence," which had actually been tested at experiment stations in the western states as early as 1914 but not released. It evidently did have commercial potential (some 6,000 acres were still listed as being planted to Florence, or Burbank's Quality, as late as 1949 in Montana, Idaho, South Dakota, and Oregon), but Burbank was inevitably accused of sharp practice. How he came to reintroduce Florence in this fashion is a matter of conjecture, and it may simply have been an accident. He could conceivably have produced an identical hybrid, or Florence seed could have found its way into wheat he had obtained for experimental purposes, been selected out, and innocently renamed as a new variety. But many experts thought not, and the fact that the same thing occurred with another Burbank wheat, named "Super," released the same year, which was identified as Jones Fife, a variety introduced by the USDA in 1893, did nothing to improve his credibility as a cereal breeder, particularly with the experiment stations.

As long before as the 1880s, he had imported seed of quinoa, an important food crop in pre-Columbian America, from Brazil. It had been offered for sale in a descriptive circular of 1897 as "a new vegetable from Brazil which is easily grown and promises to be the most valuable of any which has been introduced for years. The seed is the principal food of thousands of people in Brazil, Peru and Bolivia. It is a delicious vegetable if cooked like Spinach." This attempt to introduce it as a green vegetable was a failure, but now, in 1918, Burbank once again endeavored to promote it, this time as "a new breakfast food, a forgotten cereal of the ancient Aztecs." It did not catch on in that capacity either, but this may well be one of Burbank's ideas that is worthy of further consideration. In the high altitudes of the Andes, where the tiny seeds are ground up for flour or porridge, quinoa is said to produce more bushels per acre than perhaps any other crop.

Another plant Burbank had been working with was garlic, and in 1919 he announced the giant variety called Elephant garlic, perhaps the result of crossing the familiar *Allium sativum* with a wild garlic from Chile.

That Burbank's business had recovered after the twin debacles of the Luther Burbank Press and the Luther Burbank Company, or that he was at any rate prospering, is indicated by his gift in 1921 of $5,000 to the development of Doyle Park, presented to the city of Santa Rosa by the banker F. P. Doyle in memorial to his only son, who had died at the age of fourteen. Doyle afterward remarked that Burbank had had "a very sentimental interest in the park tract for the reason that when a young man and new to Santa Rosa . . . he did not attend church because he considered his clothes to be too shabby, so instead he spent many Sundays communing with Nature under the trees of what is now Doyle Park."

The torrent of words continued to flow. In 1921 an eight-volume condensation of the Williams books was issued by F. P. Collier under the title *How Plants Are Trained to Work for Man.* In July that year, Burton C. Bean, designated Burbank's "official biographer" on specially printed letterheads, began gathering material. After conducting interminable interviews with Burbank, he was superseded by Wilbur Hall, who was to ghost Burbank's posthumously published autobiography *The Harvest of the Years*. Bean evidently sold Hall and Elizabeth Burbank his notes and his interest in the project for $2,500 down, against a promised $10,000.* By repute a hard-drinking atheist, Hall was perhaps an unlikely candidate to be Burbank's amanuensis, but he was favored by Elizabeth Burbank, and the collaboration went forward.

Too little noticed was the most significant visitor Burbank had that year, the great Russian scientist Nikolai Ivanovich Vavilov (1887–1942). What transpired at their meeting is not recorded. Vavilov's work was to be of great importance in tracing the origins of cultivated plants, and he was well aware that much of Burbank's success was owed to his canny importations. He later noted that the success of plant breeders such as Burbank was "chiefly based upon their use, in crossing, of extensive varietal materials from foreign countries."[2] Famine conditions prevailed in the newly established USSR. It was the period of "war communism," and a *cordon sanitaire* had been imposed to seal off the revolutionary state from the rest of the world. Russian agricultural scientists were searching for new varieties—new strains of wheat, rye, and barley in particular—to alleviate the hunger of their people. Among the plant materials Vavilov obtained from Burbank may have been seed of a

* There is reason to think that Emma Burbank Beeson was paid for the use of the title "The Harvest of the Years," which she perhaps originated. See also chapter 4, note 4.

new variety of sunflower the latter had recently introduced. Improvement in this plant would have been of considerable interest to growers in Russia, where sunflowers were extensively raised for the oil from their seeds.

Vavilov would almost certainly have disagreed with his host's views on heredity. In fact, Burbank's belief in the inheritance of acquired characteristics was subsequently to be adduced in support of the theories of Stalin's protégé T. D. Lysenko, whose influence blighted Russian plant breeding in the years 1937–64 and whose supporters hounded Vavilov and many other geneticists in the USSR to their deaths as "Mendelist-Morganist" reactionaries. Lysenko's proposition that "heredity is the concentrate, as it were, of the environmental conditions assimilated by plant organisms in a series of preceding generations" sounds almost like a paraphrase of Burbank's dictum that "heredity is the sum of all past environments," and Lysenko referred to Burbank as one of "the best biologists."[3]

In December 1924, Burbank had a visitor of a very different stamp. Paramahansa Yogananda had come to America in 1920 as a delegate to the International Congress of Religious Liberals held that year in Boston under the auspices of the American Unitarian Association. He visited Santa Rosa in the course of a transcontinental lecture tour in 1924. (This seems to have been a great success, inasmuch as Yogananda was able to found his Self-Realization Fellowship in Los Angeles the following year.) He and Burbank apparently hit it off wonderfully. Yogananda tells in his autobiography of Burbank's remarking that in working with spineless cactus he had often talked to the plants "to create a vibration of love," saying, "You don't need your defensive thorns. I will protect you."[4] Evidently he also told Yogananda that his mother had appeared to him in visions and spoken to him many times since her death. For his part the swami dubbed Burbank "an American saint" and dedicated his *Autobiography of a Yogi* to him. When Burbank died, Yogananda informs us, he conducted a Vedic memorial rite in New York before a large picture of him.

He had apparently initiated Burbank into his technique of Kriya Yoga—described as a "simple, psychophysiological method by which human blood is decarbonated and recharged with oxygen" with "nothing in common with the unscientific breathing exercises taught by a number of misguided zealots"—and Burbank is quoted as saying, "I practice the technique devoutly, Swamiji." Yogananda also obtained an endorsement from him, reproduced in facsimile in the *Autobiography of a Yogi.* In a letter dated December 22, 1924, Burbank declares:

> I have examined the Yogoda system of Swami Yogananda and in my opinion it is ideal for training and harmonizing man's physical, mental, and spiritual natures. Swami's aim is to establish 'How-to-Live' schools throughout the world, wherein education will not confine itself to intellectual development alone, but also training of the body, will, and feelings.
>
> Through the Yogoda system of physical, mental, and spiritual unfoldment by simple and scientific methods of concentration and meditation, most of the complex problems of life may be solved, and peace and good-will come upon earth. The Swami's idea of right education is plain commonsense, free from all mysticism and non-practicality; otherwise it would not have my approval.
>
> I am glad to have this opportunity of heartily joining with the Swami in his appeal for international schools on the art of living which, if established, will come as near to bringing the millennium as anything with which I am acquainted.[5]

But the years had taken their toll. Despite Yogananda's Kriya Yoga, despite the outdoor life and the temperance he so vigorously championed,[6] despite his own protective hypochondria, he was an old, unwell man. The best years of his work were long behind him, in the 1890s. In the eyes of those whose opinions he most valued, he was at no very great remove from being a fraud. He was too hardheaded to have much respect for the legions of his admirers. How much enthusiasm could he, for example, have had for the birthday pageant concocted in his honor by Ada Kyle Lynch? Commencing and ending with a special "Burbank March," it assembled a troupe of unfortunate schoolchildren costumed as "flowers, fruits, buds or blossoms," or allegorically got up as Father Time, the Seasons, America, Other Nations, and what not, chanting the authoress's "descriptive poem," "Luther Burbank," and singing her unmelodious "Birthday Song" and "All Hail to Luther Burbank." Four were selected to read his life story, dressed as white, red, yellow, and pink roses. The "Birthday Song" followed, with one member of the cast carrying a large banner with a "picture of Mr. Burbank in the center, bordered with a large heart of red. Above the heart in large green letters and figures is the date of Mr. Burbank's birth, and just below the apex of the heart the number of years of the birthday being celebrated." The whole farrago took three-quarters of an hour, concluding with a Yell. It was the sort of thing to make any genuine child lover cringe, and Burbank was a genuine child lover.[7]

Who was to carry on his work after his death? This nagging question had by no means been decided. Various plans mooted to convert his gardens into an experiment station to be operated by the two universities, Stanford and Berkeley, had come to naught. In 1922 Burbank hired Will Henderson, an assistant who turned out to be especially promising and who was gradually given an unprecedented degree of responsibility. But he never committed himself to naming Henderson his successor, and the latter being very young—still in his early twenties when Burbank died—no transfer of authority ever took place.

The newspapers, dubbing him "the Wise Man of the West," continued to seek his opinions on everything under the sun, and he could never resist giving them. An interview with the San Francisco *Examiner* in 1925 elicited pronouncements on subjects ranging from current homicides to jazz. With crime he could apply his theories of hybrid nature, declaring, "We have more geniuses and more criminals than any other nation." Jazz, about which he knew nothing, he simply denounced: "I hate the very word and as for the thing itself and all it stands for, I can't bear even to think of it." SPIRIT OF JAZZ WILL DRAG RACE TO LEVEL OF SAVAGES, SAYS SAGE ran the headline.[8]

Early in 1926, the San Francisco *Bulletin* sent a young reporter named Edgar Waite to Santa Rosa "to quiz Burbank as to his theories on immortality and reincarnation." In responding, Burbank said, as quoted:

> As for Christ—well, he has been most outrageously belied. His followers, like those of many scientists and literary men who produce no real thoughts of their own, have so garbled his words and conduct that many of them no longer apply to present life. Christ was a wonderful psychologist. He was an infidel of his day because he rebelled against the prevailing religions and government. I am a lover of Christ

> as a man, and his work and all things that help humanity, but nevertheless just as he was an infidel then, I am an infidel today.

On January 22, the *Bulletin*, "shielding its sensational 'beat' against the buccaneering plagiarisms of rival papers, rent wide its pages to make space for my copyright interview," Waite later wrote vaingloriously.[9] The news that Burbank, the hero of the Sunday schools, had declared himself an infidel flashed across the United States. It is difficult now to conceive of the furor created. Letters poured in abusing him as an atheist, and he was besieged by demands that he recant his statement. In Santa Rosa women of the evangelical churches formed prayer groups to beseech God that Burbank be enabled to see the light. The Sonoma County Women's Christian Temperance Union—of which he was an honorary life member—invited him to a meeting to pray for his soul. (He communicated his regrets, and only ten of the ladies reportedly turned up.) He was compared to the celebrated agnostic Robert G. Ingersoll and to Clarence Darrow. Doubtless it was recalled that a letter from Burbank had been submitted in evidence by the defense at the famous "Monkey Trial" of John Scopes in Tennessee a few months earlier, and that Burbank had allegedly made disparaging remarks about William Jennings Bryan, the hero of the fundamentalists, in that connection.[10]

Burbank's friends sought to find an opportunity for him to answer his critics. Finally, he accepted an invitation from James Gordon, pastor of the First Congregational Church in San Francisco, to do so. He spoke there on the last Sunday in January, a day of pouring rain, and more than 2,500 people crammed into an auditorium with a seating capacity of 2,000 in order to hear him. Hundreds of them, his friend Frederick Clampett remarked, had perhaps not attended a church service for years. "There was something strangely fascinating about the man as he stood there," Clampett later wrote.

> His clear, thin voice filled that vast auditorium. Frail in form, with pale face and classic head, no man, I will venture to say, ever stood on that spot whose personality suggested such startling contrasts. It seemed to me as if a prophet had sprung to life out of the ages. Knowing his dread of public functions, his shyness and reserve, I followed his opening sentences, my own throat tight with misgiving. But I was soon made to realize that he more than measured up to the require-

ments of the occasion. As he went on with slow, almost hesitant speech, a stillness like unto death came over the great audience, and men and women hung upon every word he uttered in a kind of spell.[11]

"I love everybody! I love everything!" Burbank began. "Some people seem to make mistakes, but everything and everybody has something of value to contribute or they would not be here.

"I love humanity, which has been a constant delight to me during all my seventy-seven years of life; and I love flowers, trees, animals and all the works of Nature as they pass before us in time and space. What a joy life is when you have made a close working partnership with Nature, helping her to produce for the benefit of mankind new forms, colors and perfumes in flowers which were never known before; fruits in form, size, color and flavor never before seen on this globe; and grains of enormously increased productiveness, whose fat kernels are filled with more and better nourishment, a veritable storehouse of perfect food—new food for all the world's untold millions for all time to come.

"All things—plants, animals and men—are already in eternity traveling across the face of time, whence we know not, whither who is able to say. Let us have one world at a time and let us make the journey one of joy to our fellow passengers and just as convenient and happy for them as we can, and trust the rest as we trust life.

"Let us read the Bible without the ill-fitting colored spectacles of theology, just as we read other books, using our own judgment and reason, listening to the voice, not to the noisy babble without. Most of us possess discriminating reasoning powers. Can we use them or must we be fed by others like babes?

"I love especially to look into the deep, worshipful, liquid eyes of Bonita, my dog, whose devotion is as profound and lasting as life itself. But better yet, I love to look into the fearless, honest, trusting eyes of a child who so long has been said by theologians to be conceived and born in sin and pre-damned at birth. Do you believe all our teachers without question? I cannot. We must 'prove all things' and 'hold fast that which is good.'

"What does the Bible mean when it distinctly says, 'By their fruits ye shall know them'? Works count far more than words with those who think clearly.

"Euripides long ago said, 'Who dares not speak his free thought is a slave.' I nominated myself as an 'infidel' as a challenge to thought for those who are asleep. The word is harmless if properly used. Its stigma has been heaped upon it by unthinking people who associate it with the bogie devil and his malicious works. The devil has never concerned me, as I have always used my conscience, not the dictum of any cult.

"If my words have awakened thought in narrow bigots and petrified hypocrites, they will have done their appointed work. The universal voice of science tells us that the consequences fall upon ourselves here and now, if we misuse this wonderful body, or mind, or the all-pervading spirit of good. Why not accept these plain facts and guide our lives accordingly? We must not be deceived by blind leaders of the blind, calmly expecting to be 'saved' by anyone except by the Kingdom within ourselves. The truly honest and brave ones know that if they are to be saved it must be by their own efforts. The truth hurts for a while as do the forceps that remove an old, useless tooth, but health and happiness may be restored by the painful removal of the disturbing member.

"My mother, who lived to the ripe old age of ninety-seven years, used very often in my boyhood and youthful days to say, 'Luther, I wish you to make this world a better place to live in than it was before you lived.' I have unfailingly tried all of my own long life to live up to this standard. I was not told to believe this or that or be damned.

"I reiterate: The religion of most people is what they would like to believe, not what they do believe, and very few stop to examine its foundation underneath. The idea that a good God would send people to a burning hell is utterly damnable to me—the ravings of insanity, superstition gone to seed! I don't want to have anything to do with such a God. I am a lover of man and of Christ as a man and his work, and all things that help humanity; but nevertheless, just as he was an infidel then, I am an infidel to-day. I prefer and claim the right to worship the infinite, everlasting, almighty God of this vast universe as revealed to us gradually, step by step, by the demonstrable truths of our savior, science.

"Do you think Christ or Mohammed, Confucius, Baal or even the gods of ancient mythology are dead? Not so. Do you think Pericles, Marcus Aurelius, Moses, Shakespeare, Spinoza, Aristotle, Tolstoi, Franklin, Emerson are dead? No. Their very personality lives and will live forever in our lives and in the lives of all those who will follow us. All of them are with us to-day. No one lives who is not influenced, more or less, by these great ones according to the capacity of the cup of knowledge which they bring to these ever-flowing fountains to be filled.

"Olive Schreiner says: 'Holiness is an infinite compassion for others; greatness is to take the common things of life, and walk truly among them.

"'All things on earth have their price; and for truth we pay the dearest. We barter it for love and sympathy. The road to honor is paved with thorns; but on the path to truth at every step you set your foot down on your own heart.

"'For the little soul that cries aloud for continued personal existence for itself and its beloved, there is no help. For the soul which knows itself no more as a unit, but as a part of the Universal Unity of which the beloved also is a part, which feels within itself the throb of the Universal Life—for that soul there is no death.'"[12]

His sermon—it was nothing else—received a maximum of publicity. It was published in the press, broadcast over the radio, and cited from the pulpits of innumerable churches. Letters and telegrams by the thousand poured in. Some were abusive. "One Bryan is worth a million Burbanks to any world, and the Bible will be doing business when you and your flowers are blowing down the years," wrote an outraged fundamentalist from Abilene, Kansas. "You will be held responsible for your statement. You have set aside the Bible, made the God of the Bible a liar, made Jesus Christ an imposter. Thus you declare yourself to be a heathen. It is too bad you have so little sense," brayed another from Keokuk, Iowa. "Isn't it just splendid to get at the exact truth?" asked a New Yorker. "The physical strength of a healthy ape is three times that of a human being, and the mental strength three times that of an evolutionist." (We may, perhaps, imagine Burbank wryly remembering a remark he had made in a lecture many years before: "Yes! It is too true all men have not arrived from Monkeydom; they were consigned as freight and will be found sidetracked at some waystation.") A minister in Pasadena dismissed the horticulturist's unprofessional pronouncements on the subject of theology as "not even as valuable as John D. Rockefeller on 'How to Play Football,'" adding that he had "gathered to his banner hundreds of Christ's enemies, some of whom had been afraid to come out in the open. Unitarians, scientists, etc., who have always belittled Christ, are with him. But God is not mocked, and I tremble for them at the result, while I pity and pray for the man who is closing a great career as a poor, deluded dupe of the Devil."

On the other hand, a Baptist minister from Kansas City, who had been expelled from his church for endorsing Burbank's statement, wrote to say how he rejoiced at his new freedom of conscience, and a miner sturdily wished "Good luck to you, dear friend Burbank," saying, "let's have the truth at all costs. Self and family has quit church going till a man comes along who will give us a god of love and forget about hell fire."

Burbank had anticipated that 1926 would be a good year. Among other things, he had perfected a number of new gladioli and a new strain of Shasta daisy. He was also at work on some new roses. In an interview with *Popular Science Monthly* shortly before his last birthday, he said that he looked forward to at least five more years of productive life. The work of the horticulturist is a long-term affair. One of the new flowers he hoped to send out that season, he remarked to a reporter in March, had been under development for almost twenty years.

Then, on the evening of March 24, 1926, he suffered a heart attack. (Two nights before, walking home with his wife from a picture show, he had apparently complained of pains in his lungs.) His friend and physician Dr. Joseph Shaw was summoned and prescribed rest and quiet. For a while it seemed that Burbank was on the road to recovery. On March 30, however, he began to suffer spasmodic hiccoughs as a result of a gastrointestinal irritation, with resultant insomnia. During the week that followed, he rallied again, but the medical reports indicated "an exhausted nervous system. . . . Prognosis is guarded, somewhat grave." The slow decline continued. Burbank was worn out, unable to eat, wracked by ceaseless hiccoughing. By Saturday, April 10, he had fallen into a coma. He died without waking at thirteen minutes after midnight that Sunday morning. His last words are said to have been, "I am better now."

He was buried on the afternoon of Wednesday, April 14, under a cedar of Lebanon beside the old house. By his own choice, the tree was his only monument. The newsreel cameras were there; so was San Francisco Mayor James Rolph, Jr., later governor of California. The service was conducted by C. S. S. Dutton, pastor of the First Unitarian Church of San Francisco, who had married Luther and Elizabeth ten years before. After Dutton had spoken, Dorothy Reagan Talbot sang "Silver Threads among the Gold," a favorite song, rather than the customary hymn.

Before he died, Burbank had evidently requested that Judge Ben B. Lindsey of Denver, a noted free-thinker, deliver the funeral address—something he had apparently asked Lindsey to do several years previously. Summoned by telephone, Lindsey rushed to Santa Rosa to do so, composing his impassioned oration en route. "Luther Burbank lives," he wound up by saying. "He lives forever in the myriad fields of strengthened grain, in the new forms of fruits and flowers, and plants, and vines, and trees, and above all, the newly watered gardens of the human mind, from whence shall spring human freedom that shall drive out false and brutal gods."[13] Finally, Wilbur Hall read the eulogy that Robert Ingersoll had pronounced at the grave of his brother Ebon, which Burbank was also said to have requested.[14]

It is impossible now to say to what extent the determinedly unreligious tone of these last rites was the contribution of Hall—a professed atheist—and of Elizabeth Burbank. It was charged and denied that Burbank had recanted his "infidel" opinions on his deathbed. In death as in life, he was a focus of controversy and a handy symbol for propagandists of all sorts. The socialist *New Masses* of June 1926, for example, carried on its rear cover a lurid drawing of him threatened by the forces of superstition and reaction, with the caption "Old pious ladies railed at him—ministers preached against him—the Ku Klux Klan threatened him—because he tried to make a heaven on earth."

But Burbank was now beyond caring about any of this.

16

BURBANK'S LEGACY

Experimentation, said Burbank, was "the sole object and purpose" of his life. Yet it is not only for his achievements as a plant breeder that he deserves to be remembered. The sum of the man was far more than that. The legacy he left behind was a complex one.

Like Ford and Edison, Burbank was so integral a part of the rapidly evolving American civilization of his time that in hindsight he seems almost to have been necessary to it. (Perhaps to say that is in itself a kind of compliment. As Robert Louis Stevenson observed, "It is the mark of a good action that it appears inevitable in the retrospect.") There were other car builders, other inventors, other plant men. But to the public at large, Ford was *the* car maker, Edison *the* inventor, Burbank *the* plant wizard. There was a special interaction between these men and their age. They became symbolic figures, unlocking the gates of change and transforming society as much by force of personal example as by their output of novelties. Inevitably the revolutions they wrought were twofold: Pandora's box is not to be opened lightly.

Burbank's influence is effectively summed up by Orland E. White, professor of agricultural biology and director of the University of Virginia's Blandy Experimental Farm. In his youth, White confessed, Burbank had been one of his heroes, but as he progressed in the study of horticulture he found himself increasingly disillusioned. After he graduated, he was asked by the editor of a botanical magazine to write a review of a new edition of Harwood's *New Creations in Plant Life*. He tackled the book in an ironical vein, and "it so happened that John Burroughs saw this review and took me seriously, writing me to the effect that as an honest scientist, which he had always thought I was, I ought to know better than to help propagandize such works and such men." Burroughs had known Burbank and had once been something of an admirer himself. His name had appeared in the advertised list of life members of the Luther Burbank Society (others were Edison, Phoebe A. Hearst, John Muir, and Hugo de Vries). But by this time—around 1918—he had, like most scientists, come to associate the name of Burbank with the worst kind of horticultural quackery.

Even so, assessing Burbank more than a decade after his death, White felt that,

> Because of his notoriety or fame it has been possible to secure a favorable and special consideration for many scientific projects that otherwise would have had a deaf ear turned toward them because of ignorance in regard to the possibilities of what they might attain. . . . He aroused the imagination of many unimaginative and ignorant people so that they became interested in a subject that otherwise they probably never would have heard of.
>
> When I explain to people that I am a geneticist, it means nothing to most of them outside of a very small scientific group. If I tell them I am a follower of Mendel, the group is only slightly increased. But if I explain to them that I am crossing plants such as Burbank did, their faces light up, and many of them begin to feel quite at home and in a position to talk intelligently about a world that they know very little concerning.[1]

This was probably the feeling of most fair-minded horticultural scientists of White's generation.

Luther Burbank died a very wealthy man. When his will was filed on April 15, 1926, his estate was appraised at upward of $168,624. Elizabeth Waters Burbank was sole heir. She lived on in the old Burbank cottage for over half a century until her own death in 1977.

Burbank's last catalogue, put together with the help of Wilbur Hall, was issued posthumously in 1926. An eight-page "final new fruits bulletin" came out in 1927. Will Henderson, the only person who knew much about Burbank's ongoing experiments, was discharged after a disagreement with his widow. Stark Brothers bought the rights to the uncompleted work.

John T. Bregger, a horticulturist brought in by Stark Brothers for the purpose, made an extensive study of the hybrid fruits being tested in the Sebastopol orchards, and in the ensuing years, numerous new varieties were introduced by the company from among these. In the 1970s, Stark Brothers were still marketing eight Burbank fruits: the July Elberta peach,[2] the Flaming Gold nectarine, the Grand Prize prune, the Elephant Heart, Red Ace, Purple Flame, and Santa Rosa plums, and the Van Deman quince. The trademark *Luther Burbank* was kept registered by the company, and Mrs. Burbank was paid a royalty each year for the use of the name.[3]

Curiously, Burbank's most important single introduction remains his first. The Russet Burbank potato, a presumed sport of Burbank's seedling, is still the most widely grown potato variety in the United States. Commonly called the "Idaho potato," it has contributed incalculably to agricultural productivity. (As long ago as 1906, Hugo de Vries quoted an official USDA statement to the effect that the Burbank potato was worth $17 million a year to the country.)[4] But Burbank's plums continue to lead, too. In California, the leading plum-producing state, Beauty, Formosa, and Santa Rosa remain prominent among early varieties; Burbank, Duarte, Wickson, and Gaviota among midseason varieties; and Giant, Ace, Late Santa Rosa, and Late Duarte among late ones.

There can be small doubt that for his plums and potato alone, Burbank must be ranked among the most successful plant breeders in history. Beside them we may place the Shasta daisy, his most successful flower, whose popularity requires no embellishment with statistics. Most of his other introductions, as was virtually inevitable, have been superseded. Tastes change, and new varieties are constantly being introduced. Only the rare exception remains in the seed catalogues for more than five or ten years.

Not all of Burbank's introductions were beneficent. In 1909 he reported on the European mistletoe, *Viscum album*, which, he said, "I now have growing on some apple trees in Sebastopol, having planted the seed several years ago from imported seed which J. C. Vaughan of Chicago supplied me by request." A parasite to numerous trees and shrubs, and rated a Class B pest, *V. album* today infests an area of over sixteen square miles around Sebastopol, representing "the only known case of introduction, establishment and spread of any mistletoe from one continent to another."[5]

Burbank's far-reaching influence in promoting scientific breeding can be seen, however, in the founding of the Eddy Tree Breeding Station, later the Institute of Forest Genetics, at Placerville, California. James G. Eddy, a lumberman of Port Blakely Mill on Puget Sound, Washington, had heard of Burbank's work in breeding the Paradox and Royal walnut hybrids, and his imagination was fired by the idea of producing hybrid forest trees for timber. In 1918 he visited Burbank and asked him if he would be prepared to head a tree-breeding project. Burbank declined, pointing out that he was an old man and that the work could not come to fruition for many years. Instead he recommended Lloyd Austin, a twenty-six-year-old pomologist with the College of Agriculture of the University of California at Davis, for the job. When Eddy set up his tree-breeding station at Placerville, Austin was appointed to take charge of it, and from this research program sprang a worldwide interest in the improvement of forest trees. The project was taken over by the Forest Service of the Department of Agriculture in 1935, and since then scientists from all over the world have come to Placerville to work with the staff of the institute. Techniques for controlled pollination in pines developed there are now employed in many countries, and a number of highly successful hybrids have been developed.[6]

Another illustration is to be found in the events surrounding the passage of the first Plant Patent Law in 1930. For many years, legislation of this kind had been sought to protect the work of plant breeders from unscrupulous exploitation in the same way that the work of other inventors is safeguarded. Little was achieved, because of the ignorance of the average congressman on the subject, until Paul Stark, then chairman of the National Committee for Plant Patents, and Archibald Augustine, president of the American Association of Nurserymen, a longtime member of the plant patent committee, took up the struggle in Washington.

Stark's committee argued before Congress that lack of patent protection was bound to have a negative effect on experimentation precisely when new vistas were opening up in the field. Among those who supported the measure was Edison, who sent a telegram urging that the existing patent law be amended to include the plant breeder. But adoption of the proposal was very much in doubt, and it was feared that if it failed to pass in the current session, legislation might not be forthcoming for many years, with disastrous results for American horticulture.

One of the main opponents of the plant patent bill was Congressman Fiorello La Guardia, later famous as mayor of New York. Acting apparently on misunderstood protests he had received, La Guardia successfully blocked passage until the sponsoring congressman, Fred S. Purnell of Indiana, inquired what he thought of Luther Burbank. "I think he is one of the greatest Americans that ever lived," the New Yorker replied flamboyantly. Purnell then proceeded to read into the record a letter written by Burbank to Paul Stark shortly before his death. "I have been for years in correspondence with leading breeders, nurserymen, and federal officials and I despair of anything being done at present to secure to the plant breeder any adequate returns for his enormous outlays of energy and money," Burbank had written.

> A man can patent a mouse trap or copyright a nasty song, but if he gives to the world a new fruit that will add millions to the value of earth's annual harvests he will be fortunate if he is rewarded by so much as having his name connected with the result. Though the surface of plant experimentation has thus far been only scratched and there is so much immeasurably important work waiting to be done in this line I would hesitate to advise a young man, no matter how gifted or devoted, to adopt plant breeding as a life work until America takes some action to protest his unquestioned rights to some benefit from his achievements.

La Guardia thereupon took the floor again, announcing, "I withdraw my objection to this bill and move its adoption." This hurdle surmounted, the House of Representatives passed the measure, and the Senate soon followed suit.[7] Burbank's posthumous authority had secured for plant breeders the protection he himself had lacked. Henceforth the originator of new and distinctly different trees or plants could rely on patent protection for seventeen years (after which they passed into the public domain). The "Little Flower"—as La Guardia was called by his supporters—might not have known much about horticulture, but he knew and honored the name of Luther Burbank.

In evaluating scientific hostility to Burbank, it is worth remembering that the age was one of conflict for workers in the field of biology. In the public arena, scientists fought the larger battle with the forces of ignorance and bigotry assembled under the fundamentalist interpretation of the Old Testament. Internal to biology itself was the protracted struggle between "Mendelian-mutationalists" and "Darwinian-biometricians." It was not until the 1930s that these factions were reconciled, with Darwin's original idea of slowly cumulative variation subsuming and modifying the radical mutationalist views of De Vries and Bateson, and Mendel's work coming into full acceptance with the biometricians who had previously adhered to Galton's law of ancestral inheritance.

Of the public struggle, we might simply note that at the Scopes trial in Tennessee, which took place in 1925, William Jennings Bryan, popular hero, several times nominee for the presidency, and sometime secretary of state, had gone so far as to deny that man was a mammal. Fundamentalism was alive and kicking, and Bryan was not the only public man to try to make a virtue out of ignorance. Edgar Anderson recalls that after Vavilov's expedition to South America had discovered polyploid complexes of wild and semiwild potatoes, the USDA was stimulated to propose sending a plant explorer to obtain a collection of these for potato breeders in the United States. But the budget for the project was ridiculed in Congress by a demagogic legislator for the reason that it contained the word "polyploid," and the expedition, which was potentially of great importance, never took place.[8] The currents of anti-intellectualism ran strong. Scientists were often on the defensive and, in defending, sometimes overstepped the bounds of objectivity themselves.

And Burbank had champions in the most unacceptable quarters. In Russia, as early as 1905, Mendelism had been dubbed "clerical anti-Darwinism" by K. A. Timiriazev, a prominent plant physiologist and Marxist. Decades later Burbank's name was appropriated by T. D. Lysenko and his followers, then engaged in ruthlessly dismantling the entire structure of Russian genetic science. "Marx had spoken well of Darwin so that Darwinism was sacred. Lenin had spoken well of Timiriazev, so that his views were sacred in the second order. Timiriazev had spoken well of the American commercial plant breeder, Burbank, so that he enjoyed third-order sanctity," C. D. Darlington, one of the foremost contemporary British geneticists, wrote in 1949. Burbank was represented as the American counterpart of the Russian breeder I. V. Michurin, whose work was held sacred by Lysenkoist dogma. "Roosevelt honoured Burbank with a purple postage stamp," Darlington riposted.

> Stalin paid his homage to Michurin with a small town, Michurinsk. In both countries societies were formed to advance their work or reputation. Both men had worked for their private profit, collecting useful plants from other countries and breeding from them. Their methods did not include such precautions as are taken by scientific plant breeders, but sometimes, as in all botanical collections, useful seedlings turned up by chance from the seed set by open pollination. In these cases both Michurin and Burbank felt able (as commercial breeders usually do) to attribute the results to "scientific" crossing with particular, and often surprising, parents that happened to be growing nearby. . . . In order to support their prodigious and, by scientific standards, fraudulent claims as creators of new plants, both Burbank and Michurin revived the good old Lamarckian theory of the direct action of a changed environment in changing heredity. They put the theory into a new dress and each probably thought he had invented it.[9]

This, as any reader of the present book will be aware, is blatantly unfair to Burbank (and it is perhaps unjust to Michurin as well). But Darlington was writing in the heat of combat. Western scientists were doing their best to aid their Soviet colleagues, who were literally in danger of their lives. Arrests of Russian geneticists had begun as early as 1932. The first death sentences came in 1935, and in 1937 S. G. Levit, head of the medical genetic research institute in Moscow, was executed. In 1940, Vavilov, the outstanding scientist in his field and the man responsible for the introduction of some 25,000 experimental samples of wheat in his country, was arrested and condemned to death as a British spy. (It is believed that he died of illness and privation in a Siberian concentration camp near Saratov in 1942.) Lysenko thereupon assumed his position as president of the Lenin Academy of Agricultural Sciences and director of the Laboratory of Genetics of the Academy of Sciences.

Lest it be doubted that Western scientists themselves felt threatened by the political climate of the times, we might quote the embattled declaration of Professor H. J. Muller of Indiana University in his presidential address to the Eighth International Congress of Genetics in 1948:

> I believe that genetics has *not yet* [emphasis added] in Western countries, gone so far toward becoming an underground movement as to make . . . an eclipse necessary. Let us struggle to keep our science, in all its ramifications, connections, extensions, and even speculations, in the light of day.[10]

None of this had anything to do with Burbank. But he had been branded "unscientific" and adopted into the bargain in Russia as a Lysenkoist totem to set beside the canonized Michurin. Accordingly, in Western eyes, he became a kind of Lysenkoist by posthumous association. Darlington does admit that

> When Roosevelt made the mistake over Burbank, the United States Department of Agriculture, the National Research Council, and the Carnegie Institution of Washington did not dismiss their very able staffs of geneticists. In Russia, however, Stalin's mistake was part of a plan which included the dismissal of the geneticists and a great deal more besides.[11]

But it is odd, indeed, to equate the honoring of Burbank on a three-cent postage stamp (one of a "Famous Americans" series commemorating artists, authors, composers, poets, inventors, and scientists, the choice was not Roosevelt's personal decision in any case) with Stalin's "mistake" in permitting the ignorant and malevolent Lysenko to obliterate Soviet genetics and hound its leading representatives to death! When they turn to polemics, scientists, it seems are no more objective than the rest of us.

Had Burbank lived fifty years earlier, there can be small doubt that he would universally be regarded as the father of American horticulture. As it was, his working career too closely preceded the rise of plant genetics and scientific breeding. He was rejected as a reactionary rival by younger men whose paths he had done a great deal to ease. They did not hesitate to judge him by the most rigorous contemporary standards, forgetting that these had not yet been established when he was doing his best work. Those who knew him well gave him his due. But the great majority simply took their cues from afar: after all, one had only to glance at the Burbank puffery in the popular press to know what one was dealing with.

A hundred years ago, the need to apply scientific methods to agriculture was far from recognized. In 1874 E. W. Hilgard started the California Agricultural Experiment Station with $250. Another pioneer experiment station was opened in Connecticut in 1875—the very year Burbank set out hopefully for California with his ten seed potatoes. Although funds from the sale of government-owned land had been set aside to support agricultural and mechanical colleges in every state by the Morrill Land Grant Act of July 1862 and the Department of Agriculture was established as an independent agency the same year, progress was slow. The system of agricultural experiment stations was not formally set up until the passage of the Hatch Act in 1887, and the head of the Department of Agriculture did not obtain cabinet rank until two years later. Once under way, however, the experiment stations, in combination with the land-grant colleges, provided a formidable nationwide basis for research and experimentation. This federal support reflected the growth of the United States as an agricultural nation. In the last quarter of the nineteenth century, the number of farms in the country more than doubled. Vast tracts of virgin land were brought under the plow. During Burbank's working life, the farm population reached its all-time peak of over 28 million.

The same period corresponded with the phenomenal growth of California agriculture. And in California, where physical circumstances tended to dictate large, single-crop farming operations rather than small diversified holdings, the applications of agricultural science and engineering were particularly obvious. The development of scientific irrigation analysis and soil study under Hilgard's guidance, the increasing mechanization of farms, and the introduction of new varieties by Burbank and other breeders laid the basis for the California agribusiness empire of today.

Burbank accurately perceived the immense possibilities of the state for fruit and vegetable growing long before most. When he arrived in Santa Rosa, California was the site of the world's largest wheat field, an eighty-mile-wide strip along the banks of the San Joaquin River that produced more than 100,000 tons annually so economically that it could profitably be shipped 14,000 miles out of San Francisco to British markets. Wheat was a bonanza far bigger than gold. The venerable Don Luis Peralta (whose rancho of five leagues in the East Bay extended over much of the future sites of the cities of Berkeley and Oakland) had spoken with remarkable foresight when he told his sons, who were eager to seek their fortunes in the gold country: "Go to your ranch and raise grain, and that will be your best gold field, because we all must eat while we live." But by 1900, wheat production had declined dramatically as a consequence of overcropping and increased surpluses on world markets, and other crops were coming in: grapes, citrus, olives, tree nuts, and deciduous fruit in particular.

One concept that evolved side by side with California agribusiness was that of crops tailored to suit market tastes, mechanical picking, and whatever other requirements might be thought desirable. Burbank must be regarded as a central innovator of these techniques. In 1905, for example, he received an order from J. H. Empson, a Colorado canner, for a pea resembling French *petits pois* that would develop and ripen uniformly and be capable of mechanical harvesting. Empson had initially believed that an entirely different variety of pea was being packed in France and ordered seed of several kinds grown there. When it was planted, however, he discovered that they were precisely the same as the varieties in cultivation in America and did not actually grow any smaller. The tender little French peas that were so sought after were made possible by the cheaper labor available in Europe: peas were picked by hand there when they were half grown and could be harvested in this way off the same vines several times during the season. The French canners also artificially sweetened their peas and used copper sulfate to give them their characteristically bright green color. Empson laudably sought to avoid these latter expedients, and he applied instead to Burbank to produce a variety that would duplicate the desired qualities naturally.

When he took on the job, Burbank contracted to produce a suitable pea in six years. He tackled it solely by selection, growing two crops a year, and by February 1908 was able to deliver the result: a pea uniformly 15 percent smaller than average, sweeter than before, maturing earlier, and producing up to twice as many pounds per acre into the bargain. It derived entirely from "Admiral" seed initially supplied by Empson, reselected through six generations.[12] This was a remarkable demonstration of the possibilities of simple selection—performed by a master selector. Jones remarks:

> To any one who has worked with naturally self-fertilized plants like peas, it seems incredible that a marked improvement could be made in such a short time with the limited amount of material that Burbank could have grown along with his many other undertakings. Breeding projects with these plants usually involve the selection of many hundreds of individuals and the testing of them in progeny plots over a period of several years under the same conditions as the plants are to be grown ultimately. When superior selections are finally found and proven it takes several years to multiply seed in sufficient quantities to make a thorough test.[13]

In this case, however, the final testing was evidently left to the canners themselves, and that they were satisfied with what they got is apparent from the fact that they began to raise and market the new pea—named the Burbank Admiral—in quantity.[14]

Today there is something of a rebellion against the standardization of produce: beautiful but relatively tasteless tomatoes and glistening apples whose superb appearance belies indifferent flavor and texture, for example. Those who have lived in parts of the world where fruit and vegetables have been less thoroughly tamed are not infrequently heard to complain about the products of scientific horticulture, in which eating quality often seems to have been the last consideration served. But it would be unfair to blame Luther Burbank. His own criteria were exacting, and flavor and scent were high on the list. When he forbade his helpers to smoke or drink, it was not so much for reasons of morality as because it would dull their senses of taste and smell to do so.

Burbank was once dubbed "the only living incarnation of Emerson's philosophy," and, like the sage of Concord, he was a product of the Unitarian tradition and hence ultimately of the Enlightenment. But, also like Emerson, he paradoxically favored intuition over law. If he stood outside the scientific tradition, it was almost by choice. "Although he professes to prefer not to know the names of his species, *lest that knowledge should limit him in his efforts at hybridization* [emphasis added], he does know by their scientific names the thousands of plants with which we has worked," Shull noted. "This became very apparent when he ran rapidly through Vilmorin's catalog checking those species which they and Mr. Burbank had in common. . . . Very often when I would mention the name of a genus he would immediately give all the species he has without the slightest hesitancy. This was something of a surprise to me because Bailey has represented Mr. Burbank as being almost wholly ignorant of the species upon which he has worked."

Burbank shaped his flowers and fruit much as an artist works up his material, using methods he and he alone had evolved and could apply, because they resided, finally, not in any procedure or technique, but in his own personal aesthetics. He tuned in to the "vibrations" that he identified as "the cause of all manifestation" and proceeded accordingly. He himself believed that this was more than what was ordinarily known as intuition, that there was something of a sixth sense to it. Even those who had worked with him longest and been closest to him, he said, had been unable to duplicate what he did as a mere matter of routine, without thinking how he did it. But he envisaged that such talents would one day come to be better understood and looked forward to a time when science would "concentrate on the wonders of the mind of man and on subjects that we now consider mystical and psychic."*

In many ways, the real Burbank is a paradox. Did he, in fact, have mysterious powers? Was he the "innately high genius" Hugo de Vries perceived? Or was it rather as the Chicago seedsman Ralph Howe sardonically observed: "Yes, he was a great man; so

* There is no doubt that Burbank believed himself to be psychic. He claimed as much in an article he wrote for *Hearst's International Magazine* in 1923, averring that he, his mother, and his sister Emma all possessed telepathic powers. He also insisted that he possessed the ability to heal by laying on of hands, citing several cases in which he had done so.

was Barnum."[15] The truth would seem to lie somewhere in the middle. Burbank undoubtedly possessed extraordinary capabilities as a plant breeder. His personal charisma fascinated the day. He played the part of sage, and even of saint, in the brassy, grasping childhood of the twentieth century. But he was prepared to exaggerate to the brink of falsehood, and perhaps beyond, to bolster his reputation. He lived frugally and dealt shrewdly. Though not, apparently, much interested in money for its own sake, he made a great deal of it. He grafted boosterism to the philosophy of Thoreau.

Burbank, it seems, could live with the contradictions of his role. "There is too much striving to be consistent rather than trying to be right," he remarked in 1918. It was an observation that might have appealed to F. Scott Fitzgerald, who said some decades later, "The test of a first-rate intelligence is the ability to hold two opposed ideas in the mind at the same time, and still function."[16]

The confident scientific pronouncements and predictions of yesteryear have turned out to be only too fallible, and there is no reason to regard those of today as any more certain. When James Watson and Francis Crick unraveled the double helix of DNA in the 1950s, it was declared to be the new "central dogma" of biology that the path of genetic information from the nucleic acids in the chromosomes to the protein-synthesizing mechanism in the cells was a one-way street. Weismann's theory of the inviolability of the germ plasm was held to be proven beyond doubt. The inheritance of acquired characteristics—"soft" inheritance, as the double entendre of biological orthodoxy dubbed it, by contrast to "hard" inheritance through natural selection alone—seemed finally excluded. "The genotype and phenotype problem could now be stated in definitive terms and the last nail could be hammered into the coffin of the theory of the inheritance of acquired characters," as Mayr puts it.[17]

But such last nails have a habit of popping out again. Crick's "central dogma" no longer holds. Genes are now perceived not as discrete, immutable atoms of heredity but as interacting components of a dynamic system. They can transpose themselves to produce new combinations and replicate themselves to make normal functioning and evolutionary development simultaneously possible. New material can be inserted into DNA from outside by agents called retroviruses. "In short, a set of new themes—mobility, rearrangement, regulation, and interaction—has transformed our view of genomes from stable and linear arrays, altered piece by piece and shielded from any interaction with their products, to fluid systems with potential for rapid reorganization and extensive feedback from their own products and other sources of RNA," writes Stephen Jay Gould. "The implications for embryology and evolution are profound, and largely unexplored."[18] Over the past decade, biotechnologists have themselves begun interfering with genotypes by means of gene splicing, protoplast fusion, and other recombinant-DNA technology.[19]

Since the publication of the first edition of this book, there has been a good deal of work both in Europe and in the United States that supports the view that some form of neo-Lamarckianism must now be admitted to respectability. This is especially true of plants, which possess a much greater diversity of genetic systems than animals.[20] "The plant genome appears to be more tolerant of variation than the animal genome," write the authors of a recent paper from Stanford University's Department of Biological Sciences.

> Being unable to move, the parent plant can exert little control over the environmental nature of the sites in which its progeny will grow and develop. One strategy for survival is for the parent plant to ensure that its progeny are genetically very variable. . . . Another possibility is for the parent plant to become genetically adapted to the environment in which it is growing so that its progeny will be better fitted to survive in that environment. The question that arises is, to what extent does the environment interact with the plant genome, especially in the meristematic cells. . . . A priori one might expect long-lived perennials to be capable of generating variability in some fraction of their many meristems in order to respond to a changing environment in their particular habitat.
>
> . . . We propose that plants not only tolerate genetic change but may have "generators" of diversity as well which act to create new genomic combinations within the plant or plant population. Such generators of diversity may be activated by stress, alien chromosomes introduced by wide crosses, or other agents and serve to diversify the genetic constitution of the plant as part of the plant's strategy to adapt to a changing environment.[21]

Mutatis mutandis, Luther Burbank and Jean Baptiste Lamarck might have found this an acceptable elaboration of their beliefs.

And it is a questionable objectivity that pretends that our way of looking at Burbank and Lamarck does not depend intimately on our way of looking at their opinions. The theorist is part of the theory. The experimenter is part of the experiment. Or so the quantum physicists have taught us to think. Burbank, who personified that view in horticulture, may even now have things to teach us.

APPENDIX I

Another Mode of Species Forming*

"Burbank's most valuable contributions to science, his *Rubus* hybrids that bred true, have been largely overlooked by geneticists," wrote Donald Jones in a review in the *Journal of the New York Botanical Garden* (December 1943). "When these were first announced, Mendelian segregation was expected in all hybrids and his statements were not accepted. Now that amphidiploids are known to transmit without segregation, Burbank is not given the priority that he deserves for producing and putting these on record." Polyploidy, since understood to be one of the most important processes in the evolution of cultivated plants, seems first to have been reported in 1906, but it was not until years later that the significance of chromosome numbers became apparent. The term *amphidiploid* was introduced only in 1927. While true-breeding species hybrids may have been observed earlier—the loganberry, which originated in the garden of Judge Logan at Santa Cruz, California, for example—Burbank's priority lies in having both produced such hybrids and clearly distinguished them as "another mode of species forming." Here is his report:

The more usual concept of the formation of species is by slow variations so well known as the Darwinian theory, which though attacked from every point, still is and must always in the

* A paper read at the annual meeting of the American Breeders' Association, Columbia, Mo., January 5–8, 1909. Reprinted in *Popular Science Monthly*, September 1909.

main be accepted, for without question it gives the fundamental principles of evolution as had never been done before. Yet the boundless amount of research along these lines during the last half century has developed strong new sidelights which illuminate, and in some cases compel a slightly different view of, some of the suggestions of *the master, Darwin.*

During the period of forty years that I have been experimenting with plant life both in bleak New England and in sunny California, extensively operating on much more than four thousand five hundred distinct species of plants, including all known economic and ornamental plant forms which are grown in the open air in temperate and semi-tropic climates, as well as many of those commonly grown in greenhouses and numerous absolutely new ones not before domesticated and on a scale never before attempted by any individual or body of individuals, numerous general principles have pressed themselves forward for discussion and observation. Only one of these can be discussed at this time, and this briefly, more as a text for further observations and experiments than as anything like a full view of this highly interesting mode of species formation.

In the first place, let me say that our so-called species are only tentative bundles of plants, no two individuals of which are exactly alike, but nearly all of which quite closely resemble each other in general outside appearances and in hereditary tendencies. Yet no one can tell just what the result will be when combinations of these inherent tendencies are crossed or subjected to any other disturbing factor or factors. Like the chemist who has new elements to work with, we may predict with some degree of accuracy what the general results will be, but any definite knowledge of the results of these combinations is far more difficult, even impossible, as the life forces of plants and animals act in infinitely more new directions than can any ordinary number of combinations of chemicals.

Only a few years ago, it was generally supposed that by crossing two somewhat different species or varieties a mongrel might be produced which might, or more likely might not, surpass its parents.

The fact that crossing was *only the first step* and *that selection from the numerous variations secured in the second and a few succeeding generations was the real work* of new plant creation *had never been appreciated*; and to-day its significance is not fully understood either by breeders or even by many scientific investigators along these very lines. Old tailings are constantly being worked over at great expense of time and with small profit, while the mother lode is repudiated and neglected.

Plant breeding to be successful must be conducted like architecture. Definite plans must be carefully laid for the proposed creation; suitable materials selected with judgment, and these must be securely placed in their proper order and position. No occupation requires more accuracy, foresight and skill than does scientific plant or animal breeding.

As before noted, the first generation after a cross has been made is *usually* a more or less complete blend of all the characteristics of both parents; not only the visible characters, but *an infinite number of invisible ones* are inherent and will shape the future character and destiny of the descendants, often producing otherwise unaccountable so-called mutations, saltations or sports, the selection and perpetuation of which give to new plant creations their unique forms and often priceless values, like the Burbank potato produced thirty-six years ago and which is now grown on this western coast almost to the exclusion of all others (fourteen millions of bushels per annum, besides the vast amount grown in the eastern United States and other countries), or the Bartlett pear, Baldwin apple and navel oranges, all of which are variations selected by some keen observer. Millions of others are forever buried in oblivion for the lack of such an observer.

But in this paper I wish to call attention to a not unusual result of crossing quite distinct wild species which deserves the most careful analysis, as it seems to promise a new text for scientific investigation, especially on biometric lines. The subject was most forcibly brought to my attention twenty years ago by the singular behavior of the second-generation seedlings of raspberry-blackberry hybrids. By crossing the Siberian raspberry (*Rubus crataegifolius*) with our native trailing blackberry (*Rubus vitifolius*), a thoroughly fixed new species was summarily produced. The seedlings of this composite *Rubus* (named *Primus*), though a most perfect blend of both parents but resembling neither, never reverted either way; all the seedlings coming much more exactly like the new type than do the seedlings of any ordinary wild rubus. Many thousand plants have been raised generation after generation, all repeating themselves after the new and unique type. No botanist on earth could do otherwise than classify it if found wild as a valid new species, which it truly is, though so summarily produced by crossing.

Since the *Primus* species was originated, numerous similar cases have attracted attention, such as my now popular Phenomenal produced by crossing the Cuthbert raspberry with our native Pacific coast blackberry, and the Logan berry, both of which, though a complete blend of two such distinct species, yet reproduce from seed as truly as any wild rubus species.

I have had also growing on my grounds for some fifteen years or more hybrids of *Rubus idaeus* and *Rubus villosus*, both red and yellow varieties. All are exactly intermediate between these two very widely different species, yet both always come true intermediates from seed, generation after generation, never reverting either way.

By crossing the great African "stubble berry" (*Solanum guine[e]nse*) with our Pacific coast "rabbit weed" (*Solanum villosum*) an absolutely new species has also been produced, the fruit of which resembles in almost every particular the common blueberry (*Vaccinium Pennsylvanicum*), and while the *fruit of neither parent species is edible*, the fruit of the *newly created one is most delicious* and most abundantly produced, and the seedlings, generation after generation though produced by the million, still, all come as true to the new type as do either parent species to *their* normal type.

Still another example of this mode may be found in my experiments with opuntias. By crossing *O. tuna* with *O. vulgaris*, thousands of seedlings have been produced, all of which, in the first, second and third generations, though a well-balanced blend of the two natural species, still come as true to the newly created species as do either parent species to their own natural types.

Not only does this new mode hold true under cultivation but species are also summarily produced in a wild state by natural crossing.

The western blackcap (*Rubus occidentalis*) and the eastern red raspberry (*Rubus strigosus*) when growing contiguous, as they very commonly do in Central British America, often cross, forming an intermediate new species which sometimes sorely crowds both of the parent species, and when brought under cultivation still firmly maintains its intermediate characters, no matter how often reproduced from seed. And still further, our common "tarweed" (*Madia elegans*) with its beautiful large blossoms often crosses with *M. saliva* with its insignificant pale yellow flowers, producing a complete intermediate. I have not yet determined whether the intermediate will reproduce true from seed, but confidently expect it to do so. Similar results among wild evergreens and deciduous trees and shrubs and herbaceous plants have been frequently and forcefully brought to my attention, leaving little doubt in my own mind that the evolution of species is by more modes than some are inclined to admit.

APPENDIX II

The Training of the Human Plant*

I. The Mingling of Races

During the course of many years of investigation into the plant life of the world, creating new forms, modifying old ones, adapting others to new conditions, and blending still others, I have constantly been impressed with the similarity between the organization and development of plant and human life. While I have never lost sight of the principle of the survival of the fittest and all that it implies as an explanation of the development and progress of plant life, I have come to find in the crossing of species and in selection, wisely directed, a great and powerful instrument for the transformation of the vegetable kingdom along lines that lead constantly upward. The crossing of species is to me paramount. Upon it, wisely directed and accompanied by a rigid selection of the best and as rigid an exclusion of the poorest, rests the hope of all progress. The mere crossing of species, unaccompanied by selection, wise supervision, intelligent care, and the utmost patience, is not likely to result in marked good, and may result in vast harm. Unorganized effort is often most vicious in its tendencies.

* First published in *Century* magazine in May 1906. Reissued in book form in 1907 (New York: Century Company).

Before passing to the consideration of the adaptation of the principles of plant culture and improvement in a more or less modified form to the human being, let me lay emphasis on the opportunity now presented in the United States for observing and, if we are wise, aiding in what I think it fair to say is the grandest opportunity ever presented of developing the finest race the world has ever known out of the vast mingling of races brought here by immigration.

By statistical abstract on immigration, prepared by the Bureau of Statistics of the Department of Commerce and Labor in Washington, I find, that, in the year 1904, 752,864 immigrants came into the United States, assigned to more than fifty distinct nationalities. It will be worth while to look carefully at this list. It shows how widely separated geographically, as well as ethnologically, is the material from which we are drawing in this colossal example of the crossing of species:

Austria-Hungary, including Bohemia, Hungary, and other Austria save Poland	117,156	
Belgium	3,976	
Denmark	8,525	
France	9,406	
Germany	46,380	
Greece	11,343	
Italy	193,296	
Netherlands	4,916	
Norway	23,808	
Poland	6,715	
Rumania	7,087	
Russia	145,141	
Spain	3,996	
Sweden	27,763	
Switzerland	5,023	
Turkey in Europe *	5,669	
England	38,620	
Ireland	36,142	
Scotland	11,092	
Wales	1,730	
Europe not specified	143	
Total Europe		707,927

British North America	2,837	
Mexico	1,009	
Central America	714	
West Indies and Miguelon	10,193	
South America	1,667	
Total America		16,420
China	4,309	
Japan	14,264	
Other Asia	7,613	
Total Asia		26,186
Total Oceania		1,555
Total Africa		686
All other countries		90
Total Immigrants		752,864

* Includes Servia, Bulgaria, and Montenegro.

Study this list from any point of view. Where has there been found a broader opportunity for the working out of these underlying principles? Some of these immigrants will mate with others of their own class, notably the Jews, thus not markedly changing the current; many will unite with others of allied speech; still others marry into races wholly different from their own, while a far smaller number will perhaps find union with what we may call native stock.

But wait until two decades have passed, until there are children of age to wed, and then see, under the changed conditions, how widespread will be the mingling. So from the first the foreign nations have been pouring into this country and taking their part in this vast blending.

Now, just as the plant breeder always notices sudden changes and breaks, as well as many minor modifications, when he joins two or more plants of diverse type from widely separated quarters of the globe,—sometimes merging an absolutely wild strain with one that, long over-civilized, has largely lost virility,—and just as he finds among the descendants a plant which is likely to be stronger and better than either ancestor, so may we notice constant changes and breaks and modifications going on about us in this vast combination of races, and so may we hope for a far stronger and better race if right principles are followed, a magnificent race, far superior to any preceding it. Look at the material on which to draw! Here is the North, powerful, virile, aggressive, blended with the luxurious, ease-loving, more impetuous South. Again you have the merging of a cold phlegmatic temperament with one mercurial and volatile. Still again the union of great native mental strength, developed or undeveloped, with bodily vigor, but with inferior mind. See, too, what a vast number of environmental influences have been at work in social relations, in climate, in physical surroundings. Along with this we must observe the merging of the vicious with the good, the good with the good, the vicious with the vicious.

II. The Teachings of Nature

We are more crossed than any other nation in the history of the world, and here we meet the same results that are always seen in a much-crossed race of plants: all the worst as well as all the best qualities of each are brought out in their fullest intensities. Right here is where selective environment counts. When all the necessary crossing has been done, then comes the work of elimination, the work of refining, until we shall get an ultimate product that should be the finest race ever known. The best characteristics of the many peoples that make up this nation will show in the composite: the finished product will be the race of the future.

In my work with plants and flowers I introduce color here, shape there, size or perfume, according to the product desired. In such processes the teachings of nature are followed. Its great forces only are employed. All that has been done for plants and flowers by crossing, nature has already accomplished for the American people. By the crossings of types, strength has in one instance been secured; in another, intellectuality; in still another, moral force. Nature alone has done this. The work of man's head and hands has not yet been summoned to prescribe for the development of a race. So far a preconceived and mapped-out crossing of bloods finds no place in the making of peoples and nations. But when nature has already done its duty, and the crossing leaves a product which in the rough displays the best human attributes, all that is left to be done falls to selective environment.

But when two different plants have been crossed, that is only the beginning. It is only one step, however important; the great work lies beyond—the care, the nurture, the influence of surroundings, selection, the separation of the best from the poorest, all of which are embraced in the words I have used—selective environment.

How, then, shall the principles of plant culture have any bearing upon the development of the descendants of this mighty mingling of races?

All animal life is sensitive to environment, but of all living things the child is the most sensitive. Surroundings act upon it as the outside world acts upon the plate of the camera. Every possible influence will leave its impress upon the child, and the traits which it inherited will be overcome to a certain extent, in many cases being even more apparent than heredity. The child is like a cut diamond, its many facets receiving sharp, clear impressions not possible to a pebble, with this difference, however, that the change wrought in the child from the influences without becomes constitutional and ingrained. A child absorbs environment. It is the most susceptible thing in the world to influence, and if that force be applied rightly and constantly when the child is in its most receptive condition, the effect will be pronounced, immediate, and permanent.

Where shall we begin? Just where we begin with the plant, at the very beginning. It has been said that the way to reform a man is to begin with his grandfather. But this is only a half-truth; begin with his grandfather, but begin with the grandfather when he is a child. I find the following quoted from the great kindergartner Froebel:

> The task of education is to assist natural development toward its destined end.
>
> As the beginning gives a bias to the whole after development, so the early beginnings of education are of most importance.

While recognizing the good that has been accomplished in the early kindergarten training of children, I must enter a most earnest protest against beginning education, as we commonly use the word, at the kindergarten age. No boy or girl should see the inside of a school-house until at least ten years old. I am speaking now of the boy or girl who can be reared in the only place that is truly fit to bring up a boy or a plant—the country, the small town or the country, the nearer to nature the better. In the case of children born in the city and compelled to live there, the temptations are so great, the life so artificial, the atmosphere so like that of the hothouse, that the child must be placed in school earlier as a matter of safeguarding.

But, some one asks, How can you ever expect a boy to graduate from college or university if his education does not begin until he is ten years of age? He will be far too old.

First I answer that the curse of modern child-life in America is over-education. For the first ten years of this, the most sensitive and delicate, the most pliable life in the world, I would prepare it. The properly prepared child will make such progress that the difference in time of graduation is not likely to be noticeable; but, even if it should be a year or two later, what real difference would it make? Do we expect a normal plant to begin bearing fruit a few weeks after it is born? It must have time, ample time, to be prepared for the work before it. Above all else, the child must be a healthy animal. I do not work with diseased plants. They do not cure themselves of disease. They only spread disease among their fellows and die before their time.

III. Differentiation in Training

I wish to lay special stress upon the absurdity, not to call it by a harsher term, of running children through the same mill in a lot, with absolutely no real reference to their individuality. No two children are alike. You cannot expect them to develop alike. They are different in temperament, in tastes, in disposition, in capabilities, and yet we take them in this precious early age, when they ought to be living a life of preparation near to the heart of nature, and we stuff them, cram them, and overwork them until their poor little brains are crowded up to and beyond the danger-line. The work of breaking down the nervous systems of the children of the United States is now well under way. It is only when some one breaks absolutely away from all precedent and rule and carves out a new place in the world that any substantial progress is ever made, and seldom is this done by one whose individuality has been stifled in the schools. So it is imperative that we consider individuality in children in their training precisely as we do in cultivating plants. Some children, for example, are absolutely unfit by nature and temperament for carrying on certain studies. Take certain young girls, for example, bright in many ways, but unfitted by nature and bent, at this early age at least, for the study of arithmetic. Very early,—before the age of ten, in fact,—they are packed into a room along with from thirty to fifty others and compelled to study a branch which, at best, they should not undertake until they have reached maturer years. Can one by any possible cultivation and selection and crossing compel figs to grow on thistles or apples on a banana-tree? I have made many varied and strange plant combinations in the hope of betterment and still am at work upon others, but one cannot hope to do the impossible.

The First Ten Years

Not only would I have the child reared for the first ten years of its life in the open, in close touch with nature, a barefoot boy with all that implies for physical stamina, but should have him reared in love. But you say, How can you expect all children to be reared in love? By working with vast patience upon the great body of the people, this great mingling of races, to teach such of them as do not love their children to love them, to surround them with all the influences of love. This will not be universally accomplished to-day or to-morrow, and it may need centuries; but if we are ever to advance and to have this higher race, now is the time to begin the work, this very day. It is the part of every human being who comprehends the importance of this to bend all his energies toward the same end. Love must be at the basis of all our work for the race; not gush, not mere sentimentality, but abiding love, that which outlasts death. A man who hates plants, or is neglectful of them, or who has other interests beyond them, could no more be a successful plant-cultivator than he could turn back the tides of the ocean with his finger-tips. The thing is utterly impossible. You can never bring up a child to its best estate without love.

Be Honest with the Child

Then, again, in the successful cultivation of plants there must be absolute honesty. I mean this in no fanciful way, but in the most practical and matter-of-fact fashion. You cannot attempt to deceive nature or thwart her or be dishonest with her in any particular without her knowing it, without the consequences coming back upon your own head. Be honest with your child. Do not give him a colt for his very own, and then, when it is a three-year-old, sell it and pocket the proceeds. It does not provoke a tendency in children to follow the Golden Rule, and seldom enhances their admiration and respect for you. It is not sound business policy or fair treatment; it is not honest. Bear in mind that this child-life in these first ten years is the most sensitive thing in the world; never lose sight of that. Children respond to ten thousand subtle influences which would leave no more impression upon a plant than they would upon the sphinx. Vastly more sensitive is it than the most sensitive plant. Think of being dishonest with it!

Here let me say that the wave of public dishonesty which seems to be sweeping up over this country is chiefly due to a lack of proper training—breeding, if you will—in the formative years of life. Be dishonest with a child, whether it is your child or some other person's child—dishonest in word or look or deed, and you have started a grafter. Grafting, or stealing,—for that is the better word,—will never be taken up by a man whose formative years have been spent in an atmosphere of absolute honesty. Nor can you be dishonest with your child in thought. The child reads your motives as no other human being reads them. He sees into your own heart. The child is the purest, truest thing in the world. It is absolute truth: that's why we love children. They know instinctively whether you are true or dishonest with them in thought as well as in deed; you cannot escape it. The child may not always show its knowledge, but its judgment of you is unerring. Its life is stainless, open to receive all impressions, just as is the life of the plant, only far more pliant and responsive to influences, and to influences to which no plant is capable of being responsive. Upon the child before the age of ten we have an unparalleled opportunity to work; for nowhere else is there material so plastic.

Traits in Plants and Boys

Teach the child self-respect; train it in self-respect, just as you train a plant into better ways. No self-respecting man was ever a grafter. Make the boy understand what money means, too, what its value and importance. Do not deal it out to him lavishly, but teach him to account for it. Instil better things into him, just as a plant-breeder puts better characteristics into a plant. Above all, bear in mind repetition, repetition, the use of an influence over and over again. Keeping everlastingly at it, this is what fixes traits in plants—the constant repetition of an influence until at last it is irrevocably fixed and will not change. You cannot afford to get discouraged. You are dealing with something far more precious than any plant—the priceless soul of a child.

Keep out Fear

And, again, keep fear out that the child may grow up to the end of the first ten-year period and not learn what physical fear is. Let him alone for that, if he is a healthy normal child; he will find it and profit by it. But keep out all fear of the brutal things men have taught children about the future. I believe emphatically in religion. God made religion, and man made theology, just as God made the country, and man made the town. I have the largest sympathy for religion, and the largest contempt I am capable of for a misleading theology. Do not feed children on maudlin sentimentalism or dogmatic religion; give them nature. Let their souls drink in all that is pure and sweet. Rear them, if possible, amid pleasant surroundings. If they come into the world with souls groping in darkness, let them see and feel the light. Do not terrify them in early life with the fear of an after-world. Never was a child made more noble and good by the fear of a hell. Let nature teach them the lessons of good and proper living, combined with an abundance of well-balanced nourishment. Those children will grow to be the best men and women. Put the best in them by contact with the best outside. They will absorb it as a plant absorbs the sunshine and the dew.

IV. Sunshine, Good Air and Nourishing Food

We cannot carry a great plant-breeding test to a successful culmination at the end of a long period of years without three things, among many others, that are absolutely essential—sunshine, good air, and nourishing food.

Sunshine

Take the first, both in its literal and figurative sense—sunshine. Surround the children with every possible cheer. I do not mean to pamper them, to make them weak; they need the winds, just as the plants do, to strengthen them and to make them self-reliant. If you want your child to grow up into a sane, normal man, a good citizen, a support of the state you must keep him in the sunshine. Keep him happy. You cannot do this if you have a sour face yourself. Smiles and laughter cost nothing. Costly clothing, too fine to stand the wear and tear of a tramp in the woods or sliding down a haystack or a cellar door, are a dead weight upon your child. I believe in good clothes, good strong serviceable clothes for young children—clothes that fit and look well; for they tend to mental strength, to self-respect. But there are thousands of parents who, not having studied the tremendous problems of environmental surroundings, and having no conception of the influence of these surroundings, fail to recognize the fact that either an overdressed or a poorly dressed child is handicapped.

Do not be cross with the child; you cannot afford it. If you are cultivating a plant, developing it into something finer and nobler, you must love it, not hate it; be gentle with it, not abusive; be firm, never harsh. I give the plants upon which I am at work in a test, whether a single one or a hundred thousand, the best possible environment. So should it be with a child, if you want to develop it in right ways. Let the children have music, let them have pictures, let them have laughter, let them have a good time; not an idle time, but one full of cheerful occupation. Surround them with all the beautiful things you can. Plants should be given sun and air and the blue sky; give them to your boys and girls. I do not mean for a day or a month, but for all the years. We cannot treat a plant tenderly one day and harshly the next; they cannot stand it. Remember that you are training not only for to-day, but for all the future, for all posterity.

Fresh Air

To develop indoors, under glass, a race of men and women of the type that I believe is coming out of all this marvelous mingling of races in the United States is immeasurably absurd. There must be sunlight, but even more is needed, fresh, pure air. The injury wrought to-day to the race by keeping too young children indoors at school is beyond the power of any one to estimate. The air they breathe even under the best sanitary regulations is far too impure for their lungs. Often it is positively poisonous—a slow poison which never makes itself fully manifest until the child is a wreck. Keep the child outdoors and away from books and study. Much you can teach him, much he will teach himself all gently, without knowing it, of nature and nature's God, just as the child is taught to walk or run or play; but education in the academic sense shun as you would the plague. And the atmosphere must be pure around it in the other sense. It must be free from every kind of indelicacy or coarseness. The most dangerous man in the community is the one who would pollute the stream of a child's life. Whoever was responsible for the saying that "boys will be boys" and a young man "must sow his wild oats" was perhaps guilty of a crime.

Nourishing Food

It is impossible to apply successfully the principles of cultivation and selection of plants to human life if the human life does not, like the plant life, have proper nourishment. First of all, the child's digestion must be made sound by sufficient, simple, well-balanced food. But, you say, any one should know this. True, and most people do realize it in a certain sense; but how many realize that upon the food the child is fed in these first ten years largely depends its moral future? I once lived near a class of people who, from religious belief, excluded all meat, eggs, and milk from the dietary of their children. They fed them vegetables and the product of cereals. What result followed? The children were anemic, unable to withstand disease, quickly succumbed to illness. There were no signs of vigor; they were always low in vitality. But that was not all. They were frightfully depraved. They were not properly fed; their ration was unbalanced.* Nature rebelled; for she had not sufficient material to perfect her higher development.

* The request has often come to me to state what I thought a "well-balanced" food especially for children. We all need food which supplies the elements of *growth* and *repair* and all, both old and young, must also have foods which yield *warmth* and *energy*. Nearly all foods contain both these elements though in greatly varying proportions and usually far from the right ones for growth and health unless a variety of foods are eaten at each meal. Growing children need a greater proportion of body-building foods, such as lean meats, fish, milk, some vegetables and fruits. They are often fed *too great a proportion* of *sweet and starchy foods*. A certain proportion of these are absolutely necessary but we all know the "starch babies" by their pale, fat, flabby, characterless faces, lusterless eyes and general lack of vitality. Less starchy foods and more fresh meats with eggs, milk, some vegetables and fruits will give more vitality, a better growth, greater intelligence, better health and a better constitution, notwithstanding the belief of some of my vegetarian friends to the contrary.

Children mostly fed on sweet and farinaceous foods are also starved for the various *salts* and *mineral elements*. These must all be supplied especially to children else they will certainly become victims of an unbalanced, unnatural, premature development and a shortening of life simply from starvation. Life, the builder, must have the necessary materials or the structure must be imperfect and incomplete.

L. B.

What we want in developing a new plant, making it better in all ways than any of its kind that have preceded it, is *a splendid norm*, not anything abnormal. So we feed it from the soil, and it feeds from the air by the aid of sunlight and thus we make it a powerful aid to man. It is dependent upon good food. Upon good food for the child, well-balanced food, depends good digestion; upon good digestion, with pure air to keep the blood pure, depends the nervous system. If you have the first ten years of a boy's or a girl's life in which to make them strong and sturdy with normal nerves, splendid digestion, and unimpaired lungs, you have a healthy animal, ready for the heavier burdens of study. Preserve beyond all else as the priceless portion of a child the integrity of the nervous system. Upon this depends their success in life. With the nervous system shattered, what is life worth? Suppose you begin the education, so-called, of your child at, say, three or four, if he be unusually bright, in the kindergarten. Keep adding slowly and systematically, with what I think the devil must enjoy as a refined means of torment, to the burden day by day. Keep on "educating" him until he enters the primary school at five, and push him to the uttermost until he is ten. You have now laid broad and deep the foundation; outraged nature may be left to take care of the rest.

The integrity of your child's nervous system, no matter what any so-called educator may say, is thus impaired; he can never again be what he would have been had you taken him as the plant-cultivator takes a plant, and for these first ten precious years of his life had fitted him for the future. Nothing else is doing so much to break down the nervous systems of Americans, not even the insane rushing of maturer years, as this over-crowding and cramming of child-life before the age of ten. And the mad haste of maturer years is the legitimate result of the earlier strain.

Neither Plant nor Child to be Overfed

Nor should the child, any more than the plant, be overfed, but more especially should not be given an unbalanced ration. What happens when we overfeed a plant, especially an unbalanced ration? Its root system, its leaf system, its trunk, its whole body, is impaired. It becomes engorged. Following this, comes devitalization. It is open to attacks of disease. It will easily be assailed by fungous diseases and insect pests. It rapidly and abnormally grows onward to its death. So with a child you can easily over-feed it on an unbalanced ration, and the result will be as disastrous as in the case of the plant. The effect of such an unbalanced ration as that fed to the children in the community I have referred to was to shorten life; they developed prematurely, and died early.

Again some one says, But how can the very poor feed their children plenty of nutritious food?

I answer that the nation must protect itself. I mean by this that it is imperative, in order that the nation may rise to its full powers and accomplish its destiny, that the people who comprise this nation must be normal physically. It is imperative, in order that the nation be normal, that the plants of the nation from which it derives its life and without which the nation dies must be sound. All human life is absolutely dependent upon plant life. If the plant life be in any measure lowered through lack of nourishment, with the inevitable lack of ability to produce the best results, the nation suffers. To the extent that any portion of the people are physically mentally or morally unfit, to that extent the nation is weakened.

Do not misunderstand me: I am not advocating paternalism in any sense; far from it. But is not the human race worth as much care as the orchards, the farms, the cattle-ranges? I would so work upon this great blending of races, upon each individual factor in it, that each factor should be called upon to do its very best, be compelled to do its very best, if it was shirking responsibility. But in any great nation there must be a large number who cannot do their best, if I may use a contradictory term, who do not seem able to rise to their opportunities and their possibilities. Already you may see in our larger cities efforts in a small way to help feed the very poor. It can be done nationally as well as municipally, and it can be done so that no loss of self-respect will follow, no encouragement and fostering of poverty or laziness.

Then, too, there are the orphans and the waifs; these must be taken into account. They must have wise, sane, consistent state aid. I am opposed to all sectarian aid. I would do away with all asylums of all types for the indigent under sectarian or private control. The nation, or the commonwealth, should take care of the unfortunate. It must do this in a broad and liberal and sane manner, if we are ever to accomplish the end sought, to make this nation rise to its possibilities. Only through the nation, or State, can this work be done. It must be done for self-protection.

V. Dangers

In the immediate future, possibly within your life and mine, unquestionably within the life of this generation, what have we most to fear in America from this vast crossing of races? Not in the vicious adults who are now with us, for they can be controlled by law and force, but in the children of these adults, when they have grown and been trained to responsible age in vice and crime, lies the danger. We must begin now, to-day, the work of training these children as they come. Grant that it were possible that every boy and girl born in the United States during the next thirty years should be kept in an atmosphere of crime to the age of ten. The result would be too appalling to contemplate. As they came to adult years, vice would be rampant, crime would go unpunished, all evil would thrive, the nation would be destroyed. Now, to the extent that we leave the children of the poor and these other unfortunates,—waifs and foundlings,—to themselves and their evil surroundings, to that extent we breed peril for ourselves.

The only way to obviate this is absolutely to cut loose from all precedent and begin systematic State and National aid, not next year, or a decade from now, but to-day. Begin training these outcasts, begin the cultivation of them, if you will, much as we cultivate the plants, in order that their lives may be turned into right ways, in order that the integrity of the state may be maintained. Rightly cultivated, these children may be made a blessing to the race; trained in the wrong way, or neglected entirely, they will become a curse to the state.

Environment

Let us bring the application still nearer home.

There is not a single desirable attribute which, lacking in a plant, may not be bred into it. Choose what improvement you wish in a flower, a fruit, or a tree, and by crossing, selection, cultivation, and persistence you can fix this desirable trait irrevocably. Pick out any trait you want in your child, granted that he is a normal child,—I shall speak of the abnormal later,—be it honesty, fairness, purity, lovableness, industry, thrift, what not. By surrounding this child with sunshine from the sky and your own heart, by giving the closest communion with nature, by feeding this child well-balanced, nutritious food, by giving it all that is implied in healthful environmental influences, and by doing all in love, you can thus cultivate in the child and fix there for all its life all of these traits. Naturally not always to the full in all cases at the beginning of the work, for heredity will make itself felt first, and, as in the plant under improvement, there will be certain strong tendencies to reversion to former ancestral traits; but, in the main, with the normal child, you can give him all these traits by patiently, persistently, guiding him in these early formative years.

And, on the other side, give him foul air to breathe, keep him in a dusty factory or an unwholesome school-room or a crowded tenement up under the hot roof; keep him away from the sunshine, take away from him music and laughter and happy faces; cram his little brains with so-called knowledge, all the more deceptive and dangerous because made so apparently adaptable to his young mind; let him have vicious associates in his hours out of school, and at the age of ten you have fixed in him the opposite traits. He is on his way to the gallows. You have perhaps seen a prairie fire sweep through the tall grass across a plain. Nothing can stand before it, it must burn itself out. That is what happens when you let the weeds grow up in a child's life, and then set fire to them by wrong environment.

The Abnormal

But, some one asks, What will you do with those who are abnormal? First, I must repeat that the end will not be reached at a bound. It will take years, centuries, perhaps, to erect on this great foundation we now have in America the structure which I believe is to be built. So we must begin to-day in our own commonwealth, in our own city or town, in our own family, with ourselves. Here appears a child plainly not normal, what shall we do with him? Shall we, as some have advocated, even from Spartan days, hold that the weaklings should be destroyed? No. In cultivating plant life, while we destroy much that is unfit, we are constantly on the lookout for what has been called the abnormal, that which springs apart in new lines. How many plants are there in the world to-day that were not in one sense once abnormalities? No; it is the influence of cultivation, of selection, of surroundings, of environment, that makes the change from the abnormal to the normal. From the children we are led to call abnormal may come, under wise cultivation and training, splendid normal natures. A great force is sometimes needed to change the aspect of minerals and metals. Powerful acids, great heat, electricity, mechanical force, or some such influence, must be brought to bear upon them. Less potent influences will work a complete change in plant-life. Mild heat, sunshine, the atmosphere, and greatly diluted chemicals, will all directly affect the growth of the plant and the production of fruits and flowers. And when we come to animal life, especially in man, we find that the force or influence necessary to affect a transformation is extremely slight. This is why environment plays such an important part in the development of man.

In child-rearing, environment is equally essential with heredity. Mind you, I do not say that heredity is of no consequence. It is the great factor, and often makes environment almost powerless. When certain hereditary tendencies are almost indelibly ingrained, environment will have a hard battle to effect a change in the child; but that a change can be wrought by the surroundings we all know. The particular subject may at first be stubborn against these influences, but repeated application of the same modifying forces in succeeding generations will at last accomplish the desired object in the child as it does in the plant.

No one shall say what great results for the good of the race may not be attained in the cultivation of abnormal children, transforming them into normal ones.

The Physically Weak

So also of the physically weak. I have a plant in which I see wonderful possibilities, but it is weak. Simply because it is weak do I become discouraged and say it can never be made strong, that it would better be destroyed? Not at all; it may possess other qualities of superlative value. Even if it never becomes as robust as its fellows, it may have a tremendous influence. Because a child is a weakling, should it be put out of the way? Such a principle is monstrous. Look over the long line of the great men of the world, those who have changed history and made history, those who have helped the race upward,—poets, painters, statesmen, scientists, leaders of thought in every department,—and you will find that many of them have been physically weak. No, the theory of the ancients that the good of the state demanded the elimination of the physically weak was, perhaps, unwise. What we should do is to strengthen the weak, cultivate them as we cultivate plants, build them up, make them the very best they are capable of becoming.

The Mentally Defective

But with those who are mentally defective—ah, here is the hardest question of all!—what shall be done with them? Apparently fatally deficient, can they ever be other than a burden? In the case of plants in which all tendencies are absolutely vicious there is only one course—they must be destroyed. In the case of human beings in whom the light of reason does not burn, those who, apparently, can never be other than a burden, shall they be eliminated from the race? Go to the mother of an imbecile child and get your answer. No; here the analogy must cease. I shall not say that in the ideal state general citizenship would not gain by the absence of such classes, but where is the man who would deal with such Spartan rigor with the race? Besides all this, in the light of the great progress now being made in medical and surgical skill, who shall say what now apparently impossible cures may not be effected?

But it is as clear as sunlight that here, as in the case of plants, constant cultivation and selection will do away with all this, so that in the grander race of the future these defectives will have become permanently eliminated from the race heredity. For these helpless unfortunates, as with those who are merely unfortunate from environment, I should enlist the best and broadest state aid.

VI. Marriage of the Physically Unfit

It would, if possible, be best absolutely to prohibit in every State in the Union the marriage of the physically, mentally and morally unfit. If we take a plant which we recognize as poisonous and cross it with another which is not poisonous and thus make the wholesome plant evil, so that it menaces all who come in contact with it, this is criminal enough. But suppose we blend together two poisonous plants and make a third even more virulent, a vegetable degenerate, and set their evil descendants adrift to multiply over the earth, are we not distinct foes to the race? What, then, shall we say of two people of absolutely defined physical impairment who are allowed to marry and rear children? It is a crime against the state and every individual in the state. And if these physically degenerate are also morally degenerate, the crime becomes all the more appalling.

Cousins

While it seems clear now in the light of recent studies that the children of first cousins who have been reared under different environmental influences and who have remained separate from birth until married are not likely to be impaired either mentally, morally or physically, though the second generation will be more than likely to show retrogression, yet first cousin marriages when they have been reared under similar environment should, no doubt, be prohibited. The history of some of the royal families of Europe, where intermarrying, with its fatal results, has so long prevailed, should be sufficient though in these cases other baneful influences have no doubt added their shadow to the picture.

Ten Generations

But let us take a still closer view of the subject. Suppose it were possible to select say, a dozen normal families, the result of some one of the many blendings of these native and foreign stocks, and let them live by themselves, so far as the application of the principles I have been speaking of are concerned, though not by any means removed from the general influences of the state. Let them have, if you will, ideal conditions for working out these principles, and let them be solemnly bound to the development of these principles—what can be done?

In plant cultivation, under normal conditions, from six to ten generations are generally sufficient to fix the descendants of the parent plants in their new ways. Sufficient time in all cases must elapse so that the descendants will not revert to some former condition of inefficiency. When once stability is secured, usually, as indicated, in from six to ten generations, the plant may then be counted upon to go forward in its new life as though the old lives of its ancestors had never been. This, among plants, will be by the end of from five to ten generations, varying according to the plant's character—its pliability or stubbornness. I do not say that lack of care and nourishment thereafter will not have a demoralizing influence, for no power can prevent a plant from becoming again part wild if left to itself through many generations, but even here it will probably become wild along the lines of its new life, not by any means necessarily along ancestral lines.

If, then, we could have these twelve families under ideal conditions where these principles could be carried out unswervingly, we could accomplish more for the race in ten generations than can now be accomplished in a hundred thousand years. Ten generations of human life should be ample to fix any desired attribute. This is absolutely clear. There is neither theory nor speculation. Given the fact that the most sensitive material in all the world upon which to work is the nature of a little child, given ideal conditions under which to work upon this nature, and the end desired will as certainly come as it comes in the cultivation of the plant. There will be this difference, however, that it will be immeasurably easier to produce and fix any desired traits in the child than in the plant, though, of course, a plant may be said to be a harp with a few strings as compared with a child.

The Personal Element

But some one says, You fail to take into account the personal element, the sovereign will of the human being, its power of determining for itself.

By no means; I give full weight to this. But the most stubborn and wilful nature in the world is not that of a child. I have dealt with millions of plants, have worked with them for many years, have studied them with the deepest interest from all sides of their lives. The most stubborn living thing in this world, the most difficult to swerve, is a plant once fixed in certain habits—habits which have been intensified and have been growing stronger and stronger upon it by repetition through thousands and thousands of years. Remember that this plant has preserved its individuality all through the ages; perhaps it is one which can be traced backward through eons of time in the very rocks themselves, never having varied to any great extent in all these vast periods. Do you suppose, after all these ages of repetition, the plant does not become possessed of a will, if you so choose to call it, of unparalleled tenacity? Indeed, there are plants, like certain of the palms, so persistent that no human power has yet been able to change them. The human will is a weak thing beside the will of a plant. But see how this whole plant's lifelong stubbornness is broken simply by blending a new life with it, making, by crossing, a complete and powerful change in its life. Then when the break comes, fix it by these generations of patient supervision and selection, and the new plant sets out upon its new way never again to return to the old, its tenacious will broken and changed at last.

When it comes to so sensitive and pliable a thing as the nature of a child, the problem becomes vastly easier.

VII. Heredity—Predestination—Training

There is no such thing in the world, there never has been such a thing, as a predestined child—predestined for heaven or hell. Men have taught such things in the past, there may be now those who account for certain manifestations on this belief, just as there may be those who in the presence of some hopelessly vicious man hold to the view, whether they express it or not, of total depravity. But even total depravity never existed in a human being, never can exist in one any more than it can exist in a plant. Heredity means much, but what is heredity? Not some hideous ancestral specter forever crossing the path of a human being. Heredity is simply the sum of all the effects of all the environments of all past generations on the responsive, ever-moving life forces. There is no doubt that if a child with a vicious temper be placed in an environment of peace and quiet the temper will change. Put a boy born of gentle white parents among Indians and he will grow up like an Indian. Let the child born of criminal parents have a setting of morality, integrity, and love, and the chances are that he will not grow into a criminal, but into an upright man. I do not say, of course, that heredity will not sometimes assert itself. When the criminal instinct crops out in a person, it might appear as if environment were leveled to the ground; but in succeeding generations the effect of constant higher environment will not fail to become fixed.

Apply to the descendants of these twelve families throughout three hundred years the principles I have set forth, and the reformation and regeneration of the world, their particular world, will have been effected. Apply these principles now, to-day, not waiting for the end of these three hundred years, not waiting, indeed, for any millennium to come, but *make* the millennium, and see what splendid results will follow. Not the ample results of the larger period, to be sure, for with the human life, as with the plant life, it requires these several generations to fix new characteristics or to intensify old ones. But narrow it still more, apply these principles to a single family,—indeed, still closer, to a single child, your child it may be,—and see what the results will be.

But remember that just as there must be in plant cultivation great patience, unswerving devotion to the truth, the highest motive, absolute honesty, unchanging love, so must it be in the cultivation of a child. If it be worth while to spend ten years upon the ennoblement of a plant, be it fruit, tree, or flower, is it not worth while to spend ten years upon a child in this precious formative period, fitting it for the place it is to occupy in the world? Is not a child's life vastly more precious than the life of a plant? Under the old order of things plants kept on in their course largely uninfluenced in any new direction. The plant-breeder changes their lives to make them better than they ever were before. Here in America, in the midst of this vast crossing of species, we have an unparalleled opportunity to work upon these sensitive human natures. We may surround them with right influences. We may steady them in right ways of living. We may bring to bear upon them, just as we do upon plants, the influence of light and air, of sunshine and abundant, well-balanced food. We may give them music and laughter. We may teach them as we teach the plants to be sturdy and self-reliant. We may be honest with them, as we are obliged to be honest with plants. We may break up this cruel educational articulation which connects the child in the kindergarten with the graduate of the university while there goes on from year to year an uninterrupted system of cramming, an uninterrupted mental strain upon the child, until the integrity of its nervous system may be destroyed and its life impaired.

I may only refer to that mysterious prenatal period, and say that even here we should begin our work, throwing around the mothers of the race every possible loving, helpful, and ennobling influence; for in the doubly sacred time before the birth of a child lies, far more than we can possibly know, the hope of the future of this ideal race which is coming upon this earth if we and our descendants will it so to be.

Man has by no means reached the ultimate. The fittest has not yet arrived. In the process of elimination the weaker must fail, but the battle has changed its base from brute force to mental integrity. We now have what are popularly known as five senses, but there are men of strong minds whose reasoning has rarely been at fault and who are coldly scientific in their methods, who attest to the possibility of yet developing a sixth sense. Who is he who can say man will not develop new senses as evolution advances? Psychology is now studied in most of the higher institutions of learning throughout the country, and that study will lead to a greater knowledge of these subjects. The man of the future ages will prove a somewhat different order of being from that of the present. He may look upon us as we today look upon our ancestors.

Statistics show many things to make us pause, but, after all, the only right and proper point of view is that of the optimist. The time will come when insanity will be reduced, suicides and murders will be greatly diminished, and man will become a being of fewer mental troubles and bodily ills. Whenever you have a nation in which there is no variation, there is comparatively little insanity or crime, or exalted morality or genius. Here in America, where the variation is greatest, statistics show a greater percentage of all these variations.

As time goes on in its endless and ceaseless course, environment must crystallize the American nation; its varying elements will become unified, and the weeding-out process will, by the means indicated in this paper, by selection and environmental influences, leave the finest human product ever known. The transcendent qualities which are placed in plants will have their analogies in the noble composite, the American of the future.

VIII. Growth

Growth is a vital process—an evolution—a marshaling of vagrant unorganized forces into definite forms of beauty, harmony and utility. Growth in some form is about all that we ever take any interest in; it expresses about everything of value to us. Growth in its more simple or most marvelously complicated forms is the architect of beauty, the inspiration of poetry, the builder and sustainer of life, for life itself is only growth, an ever-changing movement toward some object or ideal. Wherever life is found, there, also, is growth in some direction. The end of growth is the beginning of decay.

Growth within, is health, content and happiness, and growing things without stimulate and enhance growth within. Whose pulses are not hastened, and who is not filled with joy when in Earth's long circling swing around our great dynamo the Sun, the point is reached where chilling, blistering frosts are exchanged for warmth and growth! When the flowers and grasses on the warm hillsides gleefully hasten up through the soft wet soil, or later when ferns, meadow rues and trilliums thrilled with awakened life, crack through and push up the loose mellow earth in small mounds—little volcanoes of growth; all these variously organized life forces are expressing themselves each in its own specific way. Each so-called species, each individual has something within itself which we call heredity—a general tendency to reproduce itself in form and habits somewhat definitely after its own kind.

New Species

Most of the ancient and even a large part of modern students of plant and animal life have held that their so-called true species never varied to any great extent, at least never varied from the standard type sufficiently to form what could scientifically be called a new species. Under this view the word heredity has had a very indefinite meaning when used in conjunction with environment; and a never-ending uncertainty has always been apparent as to their relative power in molding individual life.

Heredity and Environment

When the great rivers of life, which we now see, commenced on this planet they did not at once leap into existence with all their present complicated combinations of forces and motions; all were very insignificant; their slender courses, though simple, were devious and uncertain, at first lacking all the wonderfully varied but slowly acquired adaptations to environment that have come with the ages; all had many obstacles to overcome, many things to learn;—and for long ages were able to respond only to the more powerful or long-continued action of external forces. Many of these frail life streams in the long race down the ages were snuffed out by unfavorable surroundings, unfavorable heredity, or the combination and interaction of both; others more successful have lived to be our contemporaries and to-day the process is still unchanged.

If a race has not acquired and stored among its hereditary tendencies sufficient perseverance and adaptability to meet all the changes to which it must always be subjected by its ever-changing environment, it will be left behind and finally destroyed, outstripped by races better equipped for the fray.

IX. Environment the Architect of Heredity

Heredity is not the dark specter which some people have thought—merciless and unchangeable, the embodiment of Fate itself. This dark, pessimistic belief which tinges even the literature of to-day comes, no doubt, from the general lack of knowledge of the laws governing the interaction of these two ever-present forces of heredity and environment wherever there is life.

My own studies have led me to be assured that heredity is only the sum of all past environment, in other words environment is the architect of heredity; and I am assured of another fact: acquired characters *are* transmitted and—even further—that *all* characters which *are* transmitted have been acquired, not necessarily at once in a dynamic or visible form, but as an increasing latent force ready to appear as a tangible character when by long-continued natural or artificial repetition any specific tendency has become inherent, inbred, or "fixed," as we call it.

We may compare this sum of the life forces, which we call heredity, to the character of a sensitive plate in the camera. Outside pictures impress themselves more or less distinctly on the sensitive plate according to their position, intensity, and the number of times the plate has been exposed to the objects (environments) in the same relative position; all impressions are recorded. Old ones fade from immediate consciousness, but each has written a permanent record. Stored within heredity are all joys, sorrows, loves, hates, music, art, temples, palaces, pyramids, hovels, kings, queens, paupers, bards, prophets and philosophers, oceans, caves, volcanoes, floods, earthquakes, wars, triumphs, defeats, reverence, courage, wisdom, virtue, love and beauty, time, space, and all the mysteries of the universe. The appropriate environments will bring out and intensify all these general human hereditary experiences and quicken them again into life and action, thus modifying for good or evil character—heredity—destiny.

Repetition

Repetition is the best means of impressing any one point on the human understanding; it is also the means which we employ to train animals to do as we wish, and by just the same process we impress plant life. By repetition we fix any tendency, and the more times any unusual environment is repeated the more indelibly will the resultant tendencies be fixed in plant, animal, or man, until, if repeated often enough in any certain direction, the habits become so fixed and inherent in heredity that it will require many repetitions of an opposite nature to efface them.

Application to Child Life

What possibilities this view opens up in the culture and development of the most sensitive and most precious of all lives which ever come under our care and culture—child life!

Can we hope for normal, healthy, happy children if they are constantly in ugly environment? Are we not reasonably sure that these conditions will almost swamp a well-balanced normal heredity and utterly overthrow and destroy a weak though otherwise good one?

We are learning that child life is far more sensitive to impressions of any kind than we had ever before realized, and it is certain that this wonderful sensitiveness and ready adaptability has not as yet by any means been put to its best possible use in child culture—either in the home or the school—and though all must admire our great educational system, yet no well-informed person need be told that it is not perfect.

X. Character

We are a garrulous people and too often forget, or do not know, that the heart as well as the head should receive its full share of culture. Much of our education has been that of the parrot; children's minds are too often crowded with rules and words. Education of the intellect has its place, but is injurious, unnatural, and unbalanced unless in addition to cultivating the memory and reason we educate the heart also in the truest sense. A well-balanced character should always be the object and aim of all education. A perfect system of education can never be attained because education is preparing one for the environment expected, and conditions change with time and place. There is too much striving to be consistent rather than trying to be right. We must learn that what we call character is heredity and environment in combination, and heredity being only *stored environment* our duty and our privilege is to make the stored environment of the best quality; in this way character is not only improved in the individual but the desired qualities are added to heredity to have their influence in guiding the slightly but surely changed heredities of succeeding generations.

Success

Cold mathematical intellect unaccompanied by a heart for the philosophic, idealistic, and poetic side of nature is like a locomotive well made but of no practical value without fire and steam; a good knowledge of language, history, geography, mathematics, chemistry, botany, astronomy, geology, etc., is of some importance, but far more so is the knowledge that all true success in life depends on integrity; that health, peace, happiness, and content, all come with heartily accepting and daily living by the "Golden Rule" ; that dollars, though of great importance and value, do not necessarily make one wealthy; that a loving devotion to truth is a normal indication of physical and mental health; that hypocrisy and deceit are only forms of debility, mental imbecility and bodily disease, and that the knowledge and ability to perform useful, honest labor of any kind is of infinitely more importance and value than all the so-called "culture" of the schools, which too often turn out nervous pedantic victims of unbalanced education with plenty of words but with no intuitive ability to grasp, digest, assimilate and make use of the environment which they are compelled each day to meet and to conquer or be conquered.

Any form of education which leaves one less able to meet everyday emergencies and occurrences is unbalanced and vicious, and will lead any people to destruction.

Every child should have mud pies, grasshoppers, water-bugs, tadpoles, frogs, mud-turtles, elderberries, wild strawberries, acorns, chestnuts, trees to climb, brooks to wade in, water-lilies, woodchucks, bats, bees, butterflies, various animals to pet, hay-fields, pine-cones, rocks to roll, sand, snakes, huckleberries and hornets; and any child who has been deprived of these has been deprived of the best part of his education.

By being well acquainted with all these they come into most intimate harmony with nature, whose lessons are, of course, natural and wholesome.

A fragrant beehive or a plump, healthy hornet's nest in good running order often become object lessons of some importance. The inhabitants can give the child pointed lessons in punctuation as well as caution and some of the limitations as well as the grand possibilities of life; and by even a brief experience with a good patch of healthy nettles, the same lesson will be still further impressed upon them. And thus by each new experience with homely natural objects the child learns self-respect and also to respect the objects and forces which must be met.

XI. Fundamental Principles

"Knowledge is Power," but it requires to be combined with wisdom to become useful. The fundamental principles of education should be the subject of earnest scientific investigation, but this investigation should be broad, including not only the theatrical, wordy, memorizing, compiling methods, but should also include *all* the causes which tend to produce men and women with sane well-balanced characters.

We must learn that any person who will not accept what he knows to be truth, for the very love of truth alone, is very definitely undermining his mental integrity. It will be observed that the mind of such a person gradually stops growing, for, being constantly hedged in and cropped here and there, it soon learns to respect artificial fences more than freedom for growth. You have not been a very close observer of such men if you have not seen them shrivel, become commonplace, mean, without influence, without friends and the enthusiasm of youth and growth, like a tree covered with fungus, the foliage diseased, and the life gone out of the heart with dry rot and indelibly marked for destruction—dead, but not yet handed over to the undertaker.

The man or the woman who moves the earth, who is master rather than the victim of fate, has strong feelings well in hand—a vigilant engineer at the throttle.

Education which makes us lazier and more helpless is of no use. Leaders use the power within; it should give the best organized thought and experience of men through all the ages of the past. By it we should learn that it is not necessary to be selfish in order to succeed. If you happen to get a new idea don't build a barbed wire fence around it and label it yours. By giving your best thoughts freely others will come to you so freely that you will soon never think of fencing them in. Thoughts refuse to climb barbed wire fences to reach anybody.

By placing ourselves in harmony and coöperation with the main high potential line of human progress and welfare we receive the benefit of strong magnetic induction currents. But by placing our life energies at right angles to it we soon find ourselves on a low-feed induction current, thus losing the help and support which should be ours.

Straightforward honesty always pays better dividends than zig-zag policy. It gives one individuality, self-respect, and power to take the initiative, saving all the trouble of constant tacking to catch the popular breeze. Each human being is like a steamship, endowed with a tremendous power. The fires of life develop a pressure of steam which, well disciplined, leads to happiness for ourselves and others; or it may lead only to pain and destruction.

To guide these energies is the work of true education. Education of rules and words only for polish and public opinion is of the past. The education of the present and future is to guide these energies through wind and wave straight to the port desired. Education gives no one any new force. It can only discipline nature's energies to develop in natural and useful directions so that the voyage of life may be a useful and happy one—so that life may not be blasted or completely cut off before thought and experience have ripened into useful fruit.

When the love of truth for truth's sake—this poetic idealism, this intuitive perception, this growth from within—has been awakened and cultivated, thoughts live and are transmitted into endless forms of beauty and utility; but to receive this new growth we must cultivate a sturdy self-respect, we must break away from the mere petrified word-pictures of others and cultivate the "still small voice" within by which we become strong in individual thought and quick in action, not cropped, hedged and distorted by outward, trivial forms, fads and fancies. Every great man or woman is at heart a poet, and all must listen long to the harmonies of Nature before they can make translations from her infinite resources through their own ideals into creations of beauty in words, forms, colors, or sounds. Mathematical details are invaluable, the compilation method is beyond reproach; intellectually we may know many things, but they will never be of any great value toward a normal growth unless there is an inward awakening, an intuitive grasp, an impelling personal force which digests, assimilates and individualizes. This intuitive consciousness, combined with extensive practical knowledge and "horse sense," has always been the motive power of all those who have for all time left the human race rich with legacies of useful thought, with ripening harvests of freedom and with ever-increasing stores of wisdom and happiness. We are now standing upon the threshold of new methods and new discoveries which shall give us imperial dominion.

NOTES

Preface

1. Walter E. Howard, personal communication to the author, January 6, 1975.

2. Donald F. Jones, "Destroyed by His Friends," *Scientific Monthly* 63 (1946): 238–39.

3. Walter L. Howard, "Luther Burbank: A Victim of Hero Worship," *Chronica Botanica* 9, no. 5/6 (Winter 1945–46): 448. A part of Jones's manuscript was eventually published in the Spragg Memorial Lectures on Plant Breeding, Michigan State College, East Lansing, 1937, pp. 57–76.

One: The Sage of Santa Rosa

1. L. C. Dunn, *A Short History of Genetics* (New York: McGraw-Hill, 1965), p. 4.

2. Hugo de Vries, "A Visit to Luther Burbank," *Popular Science Monthly*, August 1905, pp. 329–47; see also his *Plant Breeding: Comments on the Experiments of Nilsson and Burbank* (Chicago: Open Court, 1907), and "Luther Burbank's Ideas on Scientific Horticulture," *Century* 73 (March 1907): 674–81. De Vries appears to have visited Burbank at least three times—in 1904, in 1906, and again in 1909. The description of his visit here is based primarily on the 1904 visit but also draws on accounts of the others.

3. E. J. Wickson, *Luther Burbank, Man, Methods and Achievements: An Appreciation*, a collection of four articles reprinted from *Sunset* magazine (San Francisco: Southern Pacific Co., 1902), no. 3, p. 282.

4. Walter L. Howard, "Luther Burbank: A Victim of Hero Worship," *Chronica Botanica* 9, no. 5/6 (Winter 1945–46): 327.

5. Ibid., p. 328.

Two: The Essential Art

1. Carl O. Sauer, *Seeds, Spades, Hearths and Herds: The Domestication of Animals and Foodstuffs*, 2nd ed. (Cambridge, Mass.: MIT Press, 1969), p. 104.

2. Ibid., p. 27.

3. Ibid., pp. 99–100.

4. See K. D. White, *Roman Farming* (London: Thames & Hudson, 1970).

5. Lynn White, Jr., makes this point very convincingly in *Medieval Technology and Social Change* (New York: Oxford University Press, 1962).

6. B. H. Slicher van Bath, quoted in Fernand Braudel, *Capitalism and Material Life, 1400–1800*, trans. Miriam Kochan (New York: Harper & Row, 1973), p. 81.

7. Quoted by J. H. P. Pafford, ed., in the Arden edition of *The Winter's Tale* (Cambridge, Mass.: Harvard University Press, 1963), pp. 169–70, from which the subsequent quotations from the play are also drawn.

8. U. P. Hedrick, *A History of Horticulture in America to 1860* (New York: Oxford University Press, 1950), p. 24.

9. Quoted in Daniel J. Boorstin, *The Americans: The Colonial Experience* (New York: Vintage Books, 1958), p. 260.

10. In the eighteenth century, Joseph Cooper of New Jersey, whom Hedrick identifies as the first man in America to undertake plant breeding as his life's work, selected potato, pea, and lettuce seed and also introduced the Cooper plum. In the 1790s, William Prince of Long Island made the first efforts at improving a fruit on a large scale, developing four varieties of greengage that held their own for many years. In 1819 John Adlum (1759–1836) introduced the Catawba grape at his experimental vineyard near Georgetown, D.C. C. M. Hovey of Cambridge, Massachusetts, introduced the famous Hovey strawberry in 1838, and Dr. W. D. Brinklé of Philadelphia (1799–1863) produced a number of new strawberries, red raspberries, pears, and other fruit. In 1842 James J. H. Gregory (1827–1910) of Marblehead, Massachusetts, who later became the purchaser of the Burbank potato, introduced the Hubbard squash. The first recorded hybridization involving native American and European grapes was Dr. William W. Valk's "Ada," the result of a cross made in 1845, and around the same time John Fisk Allen of Salem, Massachusetts, produced his Allen's Hybrid grape. In the 1850s, Ephraim W. Bull (1805–95) of Concord, Massachusetts, introduced the Concord grape, still the most important blue-black variety of the eastern United States, and in the same years, the New England breeder E. S. Rogers (1826–99) developed some forty-five seedling grapes, more than a dozen of which became well known. At Cincinnati, Ohio, Nicholas Longworth (1783–1863) introduced new varieties of strawberries, raspberries, and grapes. Jacob Moore (1835–1908) spent a small fortune and a lifetime's work on the production of new fruits, among them successful grapes, currants, and pears, and died in poverty. Robert Buist (1805–80) of Philadelphia, a prolific writer on horticultural subjects, was known for his camellias and hybrid verbenas.

11. A. W. Livingston, *Livingston and the Tomato: Being a History of Experiences in Discovering the Choice Varieties Introduced by Him, with Practical Instructions for Growers* (Columbus, Ohio: A. W. Livingston's Sons, 1893), pp. 20ff.

Three: The Winds of Biology

1. Thomas Andrew Knight, "Introductory Remarks Relative to the Objects Which the Horticultural Society Have in View," *Transactions of the Horticultural Society* (London), April 2, 1805, quoted in Walter L. Howard, "Luther Burbank: A Victim of Hero Worship," *Chronica Botanica* 9, no. 5/6 (Winter 1945–46):351.

2. H. F. Roberts, *Plant Hybridization before Mendel* (1929; reprint, New York: Hafner, 1965), p. 89.

3. Ernst Mayr, *The Growth of Biological Thought: Diversity, Evolution, and Inheritance* (Cambridge, Mass.: Harvard University Press, Belknap Press, 1982), pp. 650–51.

4. Ibid., pp. 305, 306.

5. Lamarck, *Discours d'ouverture. . . .* ed. Alfred Giard (Paris, 1907), p. 50, quoted in Madeleine Barthélmy-Madaule, *Lamarck the Mythical Precursor: A Study of the Relations between Science and Ideology* (Cambridge, Mass.: MIT Press, 1982) p. 48.

6. H. Graham Cannon, *Lamarck and Modern Genetics* (Manchester: Manchester University Press, 1959), p. 64.

7. Mayr, *Growth of Biological Thought*, p. 350.

8. Loren Eiseley, *Darwin's Century: Evolution and the Men Who Discovered It* (Garden City, N.Y.: Doubleday Anchor, 1961), p. 98. Lyell's discussion of Lamarck's theories is in his *Principles of Geology* (London: John Murray, 1830–33).

9. Cannon, *Lamarck*, p. 9.

10. Loren Eiseley, *Darwin and the Mysterious Mr. X: New Light on the Evolutionists* (New York: E. P. Dutton, 1979), p. 75.

11. Roberts, *Plant Hybridization*, pp. 96–97.

12. *The Autobiography of Charles Darwin*, ed. Nora Barlow (New York: W. W. Norton, 1969), p. 120.

13. Mayr, *Growth of Biological Thought*, p. 268.

14. Fisher is quoted in Mayr, *Growth of Biological Thought*, p. 547; for Mayr's comment see ibid., p. 549.

15. Gavin de Beer, *Charles Darwin: A Scientific Biography* (Garden City, N.Y.: Doubleday, 1965), p. 177.

16. Cannon, *Lamarck*, p. 71.

17. Quoted in Eric F. Goldman, *Rendezvous with Destiny* (New York: Alfred A. Knopf, 1958), p. 92.

18. *Luther Burbank: His Methods and Discoveries and Their Practical Application*, ed. Henry Smith Williams et al. (Santa Rosa: Luther Burbank Press, 1915), 2:70.

19. Luther Burbank, with Wilbur Hall, *The Harvest of the Years* (Boston: Houghton Mifflin, 1927), p. 22.

20. Eiseley, *Darwin's Century*, p. 210.

21. Ibid., pp. 216–17.

22. C. D. Darlington, "Purpose and Particles in the Study of Heredity," in *Science, Medicine and History*, ed. E. A. Underwood (New York: Oxford University Press, 1953), 2:474.

23. Charles Darwin, *The Origin of Species*, 6th ed. (1872; London: Everyman's Library, 1956), p. 202, quoted in Cannon, *Lamarck*, p. 37.

24. Cannon, *Lamarck*, p. 35.

25. Barthélmy-Madaule, *Lamarck*, p. 81; see also Genesis 30 : 37–39, which illustrates one early view of the transmission of characters.

26. Hugo de Vries, *Intracellular Pangenesis*, trans. C. S. Gager (Chicago: Open Court, 1910), Introduction.

27. August Weismann, "On the Supposed Botanical Proof of the Transmission of Acquired Characters," in *Essays upon Heredity and Kindred Biological Problems*, trans. E. S. Schönland (Oxford: Clarendon Press, 1889), p. 387, quoted in Barthélmy-Madaule, *Lamarck*, p. 82.

28. E. E. Just, "Unsolved Problems of General Biology," *Physio. Zool.* 13 (1940) : 123, quoted in Cannon, *Lamarck*, pp. 68–69.

29. See, for example, E. J. Steele, *Somatic Selection and Adaptive Evolution: On the Inheritance of Acquired Characters*, 2nd ed. (Chicago: University of Chicago Press, 1981), passim; R. M. Gorczynski and E. J. Steele, "Simultaneous yet Independent Inheritance of Somatically Acquired Tolerance to Two Distinct H–2 Antigenic Determinants in Mice," *Nature* 289 (February 19, 1981) : 678–81; and Colin Tudge, "Lamarck Lives—in the Immune System," *New Scientist* 89 (February 19, 1981) : 483–85.

30. *Luther Burbank: His Methods and Discoveries*, 2 : 96–98.

31. C. D. Darlington and K. Mather, *The Elements of Genetics* (New York: Schocken Books, 1969), p. 263.

32. Quoted by David Starr Jordan, letter to Luther Burbank, July 29, 1904, Luther Burbank papers, Library of Congress, box 6.

33. Mayr, *Growth of Biological Thought*, p. 547.

34. Steele, *Somatic Selection and Adaptive Evolution*, p. 90.

Four: New England Rock

1. Henry Adams, *The Education of Henry Adams* (Boston: Houghton Mifflin, 1918), p. 53.

2. Ibid. In his introduction, D. W. Brogan observes that "there were no legitimate descendants in the male line of Washington, Franklin, Jefferson, the only founders of the Republic to be compared with John Adams" (p. viii).

3. The hour of Burbank's birth is given by the astrologer Marc Edmund Jones, but there is no indication of how Jones came by it and no independent corroboration. It was presumably obtained from Burbank either by Jones or by someone else.

4. Emma Burbank Beeson, *The Early Life and Letters of Luther Burbank* (San Francisco: Harr Wagner, 1927); originally published in May 1926 as *The Harvest of the Years: Early Life and Letters of Luther Burbank*. Presumably the title was changed because Wilbur Hall wanted to use "The Harvest of the Years" for his "autobiography" of Burbank. All quotations here attributed to Emma are from the 1927 edition.

5. Among the records in Lunenburg town hall, Walter Howard found a yellowing manuscript with the title "A History of the Town of Lunenburg in Massa-

chusetts from the Original Grant December 7, 1719, to January 1, 1866," by George A. Cunningham. This gives some further details of the Burbank family, including a list of the children born in Lancaster. There were:

1. Susan E., born 2 September, 1822, died 20 July, 1825, aged 3 years.
2. Sarah M., born 21 February, 1826, married November, 1846, A. F. Kidder of Lancaster, where they both died, leaving two children. 1. Marcia L. and 2, Lizzie.
3. Hannah E., born 5 April, 1828, died 23 March, 1843, aged 15 years.
4. George W., born 17 November, 1829, married Apphie R. Blake. Settled in California.
5. Lucy A., born April, 1831, died 29 May, 1848, aged 17 years, 1 month and 25 days.
6. Hosea Herbert, born 13 October, 1834, married 7 November, 1860, Lizzie H. Anderson, born Grafton, 1833. Lived in Westfield; one son, Henry.
7. Eliza Jenny, born 17 April, 1836, married George Varnum Ball.
8. David B., born 6 August, 1838, married 20 August, 1864, Paulina V. Ball, born 17 August, 1838, daughter of Rev. Hosea and Sarah (Helmes) Ball. They are—1874—living in California.

By second [?] wife Olive:

9. Luther, born 7 March, 1849.
10. Alfred Walton, born 2 February, 1852.
11. Emma Louisa, born 20, July 1854.

The list omits all mention of Samuel Burbank's second wife, Mary Ann, and her two children, and of Olive's first two babies, who also died in infancy.

6. Luther Burbank, with Wilbur Hall, *The Harvest of the Years* (Boston: Houghton Mifflin, 1927), p. 2.

7. John Tebbel, *The Media in America* (New York: Thomas Y. Crowell, 1975), passim.

8. Luther Burbank, letter to E. J. Wickson, November 8, 1907. Wickson papers, Library of the University of California, Davis.

9. Burbank, *Harvest of the Years*, p. 7.

10. Luther Burbank, *The Training of the Human Plant* (New York: Century Co., 1907), p. 91.

11. Burbank, *Harvest of the Years*, pp. 5–6.

12. The Lancaster Academy was a sort of superior prep school for Harvard and Yale. The course of study was comparatively stiff. As listed in the school's catalogue for 1866–67, it included:

First Year.	Eaton's *Arithmetic*, Greene's *Grammar*, Losing's *U.S. History*.
Second Year.	Robinson's *Algebra*, Harkness' *Latin*, Grammar & Reader, Caesar's *Commentaries* begun, Carter's *Physical Geography*, Quackenbos' *Philosophy*, Worcester's *History*, Otto's *French Grammar*, Fulton and Eastman's *Book-keeping*.
Third Year.	Robinson's *Geometry*, Caesar's *Commentaries* finished, Cicero's *Orations*, Crosby's *Greek Grammar and Lessons*, *Anabasis* of Xenophon begun, Sewell's *History of Greece*, Corson's *Soirées Littéraires*, Comer's *Book-keeping*, Cutter's *Physiology*.
Fourth Year.	Arithmetic, Algebra and Geometry reviewed, *Aeneid* and *Georgics* of Virgil, *Anabasis* of Xenophon finished, Three Books of Homer's *Iliad*,

Modern Series of French Plays, Surveying with field-practice, Blockheart's *Chemistry.*

Fifth Year. (Corresponding to Freshman Year in Harvard College.) Pierce's *Algebra* and *Geometry, Odes* and *Epodes* of Horace, Lincoln's *Livy* (Books XII and XIII), *Memorabilia* of Xenophon, *Odyssey* of Homer (Three Books), Felton's *Greek Historians, Histoire Grecque par Dury.* Regular reading classes continue throughout the 1st, 2nd and 3rd years, and exercises in Composition throughout the course.

In the catalogue for 1867–68, in which Burbank's name also appears in the list of students, Rolfe and Gillett's *Natural Philosophy* and *Chemistry* and Guyot's *Earth and Man* are added to the above list of textbooks. Presumably, more emphasis was given to science as the years went by. Burbank's name does not appear in the catalogues subsequent to 1867–68, however. The academy, which dated from 1815, closed in 1873.

13. Frederick W. Clampett, *Luther Burbank, "Our Beloved Infidel"* (New York: Macmillan, 1926), pp. 19–20.

14. "Up Fitchburg Way," MS reminiscence, Luther Burbank papers, Library of Congress.

Five: Burbank's Seedling

1. Luther Burbank, with Wilbur Hall, *The Harvest of the Years* (Boston: Houghton Mifflin, 1927), p. 10.

2. Ibid., p. 11.

3. Ibid., p. 12.

4. Donald F. Jones, "The Life and Work of Luther Burbank," unpublished manuscript, quoted by permission of the Connecticut Agricultural Experiment Station, p. 24.

5. H. G. Baker, *Plants and Civilization*, 2nd ed. (Belmont, Calif.: Wadsworth Publishing Co., 1970). See also Redcliffe N. Salaman, *The History and Social Influence of the Potato* (New York: Cambridge University Press, 1949). Salaman devotes almost 700 pages to the history of what he calls "this enigmatical root."

6. See J. C. Walker, "Genetics and Plant Pathology," in *Genetics in the Twentieth Century*, ed. L C. Dunn (New York: Macmillan, 1951).

7. U. P. Hedrick, *A History of Horticulture in America to 1860* (New York: Oxford University Press, 1950), p. 44.

8. Walter L. Howard, "Luther Burbank: A Victim of Hero Worship," *Chronica Botanica* 9, no. 5/6 (Winter 1945–46): 477.

9. Information supplied by the U.S. Department of Agriculture, Agricultural Research Service, Beltsville, Maryland.

10. Vernon L. Parrington, *Main Currents in American Thought* (New York: Harcourt, Brace, 1930), 3: 8.

11. *Luther Burbank: His Methods and Discoveries and Their Practical Application*, ed. Henry Smith Williams et al. (Santa Rosa: Luther Burbank Press, 1915), 12: 61.

12. Henry James, *The Bostonians* (New York: Modern Library, 1956), p. 343.

13. Parrington, *Main Currents*, 3: 11.

Six: "The Chosen Spot of All This Earth"

1. Luther Burbank, with Wilbur Hall, *The Harvest of the Years* (Boston: Houghton Mifflin, 1927), p. 31.

Seven: Plums from Yokohama

1. *Luther Burbank: His Methods and Discoveries and Their Practical Application*, ed. Henry Smith Williams et al. (Santa Rosa: Luther Burbank Press, 1915), 12:72–74.

2. Luther Burbank, with Wilbur Hall, *The Harvest of the Years* (Boston: Houghton Mifflin, 1927), p. 39.

3. Walter L. Howard, "Luther Burbank: A Victim of Hero Worship," *Chronica Botanica* 9, no. 5/6 (Winter 1945–46):336.

4. Ibid., pp. 336–37.

5. Ibid., p. 339.

6. *Luther Burbank: His Methods and Discoveries*, ed. Williams et al., quoted in Howard, "Luther Burbank: A Victim of Hero Worship," p. 340. Howard's italics.

7. Luther Burbank, letter to H. E. van Deman, August 22, 1888. Howard papers, Bancroft Library, University of California, Berkeley.

8. H. M. Butterfield, "History of California Plum and Prune Production," the third article in a series on the deciduous fruit industry in California published in *The Blue Anchor*, the official publication of the California Fruit Exchange, vols. 14 and 15, Sacramento, August 1937 to April 1938. Most of the details given here on the early history of plums and prunes in California are from this source. Butterfield apparently had access to certain notes on the subject kept by Burbank, which were supplied to him by John T. Bregger, a horticulturist employed by Stark Brothers to make a study of the Sebastopol experimental orchard after Burbank's death, but his dating of Burbank's introductions often seems faulty and at odds with the latter's catalogues.

9. *Luther Burbank: His Methods and Discoveries*, ed. Williams et al., 12:108.

Eight: "New Creations"

1. Burbank's meeting with Helen Coleman cannot be precisely dated. Howard states that it took place "on one of his transcontinental trips," and this seems to have been the only one in the years in question.

2. William Allen White, *The Autobiography of William Allen White* (New York: Macmillan, 1946), p. 176.

3. G. H. Shull, letter to W. L. Howard, February 22, 1940. Howard papers, Bancroft Library, University of California, Berkeley.

4. Walter L. Howard, "Luther Burbank: A Victim of Hero Worship," *Chronica Botanica* 9, no. 5/6 (Winter 1945–46):377.

5. Ibid., pp. 322, 375.

6. Donald F. Jones, "The Life and Work of Luther Burbank," unpublished manuscript, quoted by permission of the Connecticut Agricultural Experiment Station, p. 178.

7. *Luther Burbank: His Methods and Discoveries and Their Practical Application*, ed. Henry Smith Williams et al. (Santa Rosa: Luther Burbank Press, 1915), 2:138–43.

8. Jones, "Life and Work of Luther Burbank," p. 117.

9. Edwin G. Conklin, "A Generation's Progress in the Study of Evolution," *Science* 80, no. 2,068 (August 17, 1934).

10. Andrew J. Coe of Meriden, Connecticut, bought the Mammoth Japan Chestnut and the hybrid raspberry "S.S.–8940" for $300 each; John Lewis Childs of Floral Park, New York, took the Santa Rosa quince (he promptly renamed it the "Childs") for $800, the plums Delaware, Shipper, and Juicy for $500 each, the Golden Mayberry for $800, the Primus berry for $600, the hybrid berry "V.C.–16407" for $800, a pair of dewberries for $200 each, the Eureka raspberry for $300, the "Sugar Hybrid" raspberry for $400, the "Dwarf Pardalium" [*sic*] lily for $500, and some lesser items besides; A. Blanc & Co. of Philadelphia, purchased the whole stock of Burbank's gladiolus "California strain" for an unknown amount; and the Sunset Seed and Plant Company of San Francisco bought the seedling roses "M.–11,120" and "M.–19,928" for $300 each.

11. Dickson Terry, *The Stark Story* (St. Louis: Missouri Historical Society, 1966), p. 32.

Nine: Burbank the Wizard

1. Stock and control of the "Giant" prune ("A.P.–90" of 1893) were offered for $2,500. Not having found a buyer the previous year, Burbank had evidently had this reproduced for marketing himself, and there were now several thousand yearling trees "which are growing east of the Mississippi, so that any Eastern purchaser will have a good start *this season*." If it was not sold by September 1, he said, he would introduce it to the general trade himself. Also offered were the "Prolific" plum ("J.–3,972" of 1893), now priced at $500 as against $300 the year before, the "Doris" plum—Stark Brothers had already purchased a half-interest in this, and the other half was offered for $300—and the "Honey" prune, another seedling of the French prune, priced at $300. A quince, "No. 80," stated to be a seedling of Rea's Mammoth, was offered for $600.

Besides the "Iceberg" white blackberry, the catalogue listed the "Humboldt" blackberry-raspberry hybrid ("V. C. 18,234" of 1893), now offered for $850 as opposed to $800 the previous year, and "Rubus Capensis," a novelty introduced, Burbank said, "by way of New Zealand from South Africa," priced at $300.

New Creations of 1894 also listed three named varieties of clematis, "Snowdrift" at $300, "Ostrich Plume" at $250, and "Waverly" at $200. These hybrids between *C. coccinea* and *C. crispa*, which he described as "a *New Race* of Clematis," were subsequently introduced by J. C. Vaughan of Chicago. Other clematis offered were about "a dozen new *double* seedlings of various forms and colors, and some single ones with *largest flowers* and unusual colors and habits of growth" at between $100 and $300 each.

His new calla, "Snowflake," is listed again, still at $2,000, and another variety, called "Fragrance" because of its unusual scent, was mentioned. Finally, the

"Peachblow" rose, which had made its appearance the previous year, was offered once more—fifty large bushes for $300. The Sunset Seed Company of San Francisco bought it in 1895, along with two other roses, "Coquito" and "Pink Pet."

2. E. J. Wickson, *California Nurserymen and the Plant Industry: 1850–1910* (Los Angeles: California Association of Nurserymen, 1921).

3. Donald F. Jones, "The Life and Work of Luther Burbank," unpublished manuscript, quoted by permission of the Connecticut Agricultural Experiment Station, p. 63.

4. Ibid., p. 113.

5. Walter L. Howard, *Luther Burbank's Plant Contributions*, University of California College of Agriculture, Agricultural Experiment Station, Berkeley, Bulletin no. 691 (March 1945), p. 95.

6. Walter L. Howard, "Luther Burbank: A Victim of Hero Worship," *Chronica Botanica* 9, no. 5/6 (Winter 1945–46): 319.

7. Luther Burbank, *Partner of Nature*, ed. Wilbur Hall (New York: D. Appleton-Century Co., 1939), p. 169.

8. Three new plums, "Apple," "America," and "Chalco"; a new prune, "Pearl"; a new rose, "Santa Rosa," of the same parentage as the Burbank rose; and a new calla, "Fragrance," which had been mentioned in the 1894 catalogue but not then offered for sale, were announced. The Paradox and Royal walnuts were also advertised. Since control had not been purchased, Burbank now introduced them himself: Paradox nuts at 50 cents apiece, Royal at 75 cents. One-year-old Royal seedlings were offered for a dollar apiece. Of the new plums, the Chalco was destined to be extensively planted, though it never became an important market variety. Apple did well in South Africa. The Pearl prune, a seedling of the French prune, pollen parent unknown, turned out to be a plum of the finest quality. It was unsuccessful as a prune because it proved difficult to dry in the open.

Ten: Fruits and Penalties of Fame

1. Donald F. Jones, "The Life and Work of Luther Burbank," unpublished manuscript, quoted by permission of the Connecticut Agricultural Experiment Station, p. 27.

2. Walter L. Howard, *Luther Burbank's Plant Contributions*, University of California College of Agriculture, Agricultural Experiment Station, Berkeley, Bulletin no. 691 (March 1945), p. 79.

3. Burbank to E. J. Wickson, August 20, 1901. This letter and other Wickson correspondence referred to in this chapter are to be found in the Wickson papers, Library of the University of California, Davis.

4. Edwin G. Conklin, "A Generation's Progress in the Study of Evolution," *Science* 80, no. 2,068 (August 17, 1934).

5. S. F. Lieb to E. J. Wickson, October 11, 1901.

6. E. J. Wickson, *Luther Burbank, Man, Methods and Achievements: An Appreciation* (San Francisco: Southern Pacific Co., 1902), pp. 9–11.

7. Jones, "Life and Work of Luther Burbank," p. 29.

8. Burbank to E. J. Wickson, February 14, 1902.

9. *Pacific Rural Press*, July 6, 1901.

10. H. E. V. Pickstone to W. L. Howard, May 18, 1938. Howard papers, Bancroft Library, University of California, Berkeley.

11. In 1942 a friend of Howard's found May Maye living in a shack on the beach at Santa Barbara, "desperately poor, half blind," in a room so full of books she could scarcely get the door open. Howard wrote to her to inquire about the Luther Burbank Company and the Luther Burbank Society. She replied, in part, "Luther Burbank was an honest man, and any frauds that they got up to were not of his doing, and he was more of a victim to them than any of the patrons who lost money, for he lost much more, the work of a lifetime, his faith in the honesty of those who came pretending to help him; and he died a broken hearted man. If you write of Luther Burbank's *work*, there are many places prior to the forming of this dishonest Company where you will find about it; and if you write an authentic life of the man, you will find that too, before he was old, and easily imposed on." Howard papers, Bancroft Library, University of California, Berkeley.

12. These opinions were offered in response to questionnaires sent out by Howard. Howard papers, Bancroft Library, University of California, Berkeley.

13. Carnegie Institution of Washington, trustees' minutes, December 12, 1905.

14. David Fairchild, *The World Was My Garden* (New York: Charles Scribner's Sons, 1939), pp. 264–65. The remark concerning Harwood's book is from a letter to Howard dated January 21, 1938. Howard papers, Bancroft Library, University of California, Berkeley.

Eleven: Enter George Shull

1. See E. J. Wickson, *California Nurserymen and the Plant Industry: 1850–1910* (Los Angeles: California Association of Nurserymen, 1921).

2. See Harry M. Butterfield, *A History of Subtropical Fruits and Nuts in California* (Berkeley: University of California, Division of Agricultural Sciences, 1963).

3. See Henry W. Kruckeberg, *George Christian Roeding, 1868–1928: A Tribute* (Los Angeles: California Association of Nurserymen, 1921).

4. Harwood says $3,500, but this conflicts with a memorandum apparently typed by Burbank himself, a copy of which is in the author's possession.

5. Carnegie Institution of Washington, Year Book No. 5 (1906), p. 24.

6. Bentley Glass, "The Strange Encounter of Luther Burbank and George Harrison Shull," *Proceedings of the American Philosophical Society* 124 (April 1980): 136.

7. Luther Burbank, *Partner of Nature*, ed. Wilbur Hall (New York: D. Appleton-Century Co., 1939), p. 194.

8. George Shull, letter to W. L. Howard, February 22, 1940. Howard papers, Bancroft Library, University of California, Berkeley.

9. Ibid.

10. W. C. Williams, MS reminiscence of Luther Burbank, quoted by permission of the director, Bancroft Library, University of California, Berkeley.

Twelve: Burbankitis

1. Francis Galton, "Hereditary Talent and Character," *Macmillan's Magazine* 12 (1865):165. Quoted in Kenneth M. Ludmerer, *Genetics and American Society: A Historical Appraisal* (Baltimore: Johns Hopkins University Press, 1972), p. 10.

2. Ernst Mayr, *The Growth of Biological Thought: Diversity, Evolution, and Inheritance* (Cambridge, Mass.: Harvard University Press, Belknap Press, 1982), p. 696.

3. Richard Hofstadter, *Social Darwinism in American Thought* (Boston: Beacon Press, 1944), p. 161.

4. Mark H. Haller, *Eugenics: Hereditarian Attitudes in American Thought* (New Brunswick, N.J.: Rutgers University Press, 1963), pp. 62–63.

5. Charles B. Davenport, "Report of Committee on Eugenics," *American Breeders' Magazine* 1 (1910):129. Quoted in Ludmerer, *Genetics and American Society*, p. 8.

6. Ludmerer, *Genetics and American Society*, p. 60.

7. "The gifts of potentiality, unit characters of the germ-plasm, are not equally shared by all people of the same race," says David Starr Jordan, *War and the Breed: The Relation of War to the Downfall of Nations* (Boston: Beacon Press, 1915), p. 33.

8. Luther Burbank, *The Training of the Human Plant* (New York: Century Co., 1907), pp. 68, 82.

9. Information supplied by Dr. F. O. Butler, medical director and superintendent of the Sonoma State Home, a prominent spokesman for sterilization, in a letter to Walter L. Howard, June 20, 1940. Howard papers, Bancroft Library, University of California, Berkeley.

10. Ludmerer, *Genetics and American Society*, pp. 125–26.

11. David Starr Jordan, *The Blood of the Nation: A Study of the Decay of Races Through the Survival of the Unfit* (Boston: American Unitarian Association, 1902), p. 7.

12. "The philosophy of Social Darwinism . . . is nowhere more compactly expressed than by General Friedrich von Bernhardi in his *Deutschland und der Nachste Krieg*," notes David Starr Jordan (*War and the Breed*, p. 92).

13. David Starr Jordan, *War and the Breed*, p. 91.

14. David Starr Jordan, *The Philosophy of Despair* (San Francisco: Paul Elder and Morgan Shephard, 1902), p. 18.

15. Quoted in Catherine Drinker Bowen, *Yankee from Olympus: Justice Holmes and His Family* (Boston: Little, Brown and Co., 1944), p. 375.

16. Quoted in Hofstadter, *Social Darwinism*, p. 165.

17. The Santa Rosa *Daily Democrat*, February 9, 1886, lists Burbank as one of those who formed a committee of local exclusionists. The letter from Burbank to Geary, dated May 1, 1892, is to be found in a family scrapbook on microfilm at the Sonoma County Library, Santa Rosa.

18. For example, in *Eugenics, Euthenics and Love—How They Go Hand in Hand*, ed. Henry Smith Williams (Santa Rosa: Luther Burbank Society, 1914) and

My Beliefs (New York: Avondale Press, n.d.), a booklet published after Burbank's death. It is impossible now to determine to what extent these actually represent Burbank's own views.

19. Burbank, *Training of the Human Plant*, pp. 71–72, 20.

20. Ernest Braunton, letter to E. J. Wickson, August 28, 1906. Wickson papers, Library of the University of California, Davis.

21. E. J. Wickson, letter to Ernest Braunton, September 9, 1906. Wickson papers, Library of the University of California, Davis.

22. E. J. Wickson, *Luther Burbank, Man, Methods and Achievements: An Appreciation* (San Francisco: Southern Pacific Co., 1902), p. 15.

23. Patrick O'Mara, "Luther Burbank: A Short Review of His Work in Plant Hybridization and Brief Comparison with Other Hybridizers," *Florist's Exchange*, October 20, 1906.

24. Paul C. Mangelsdorf, "Hybrid Corn: Its Genetic Basis and Its Significance in Human Affairs," in *Genetics in the Twentieth Century*, ed. L. C. Dunn (New York: Macmillan, 1951).

25. Bentley Glass, "The Strange Encounter of Luther Burbank and George Harrison Shull," *Proceedings of the American Philosophical Society* 124 (April 1980): 138.

26. R. S. Woodward, letter to G. H. Shull, September 5, 1906. Files of the Carnegie Institution.

27. E. Carleton MacDowell's notes for a history of Cold Spring Harbor, quoted in Glass, "Strange Encounter," p. 136.

Thirteen: The Luther Burbank Press

1. W. E. Castle, "The Beginnings of Mendelism in America," in *Genetics in the Twentieth Century*, ed. L. C. Dunn (New York: Macmillan, 1951), p. 65.

2. Charles B. Heiser, Jr., *Nightshades: The Paradoxical Plants* (San Francisco: W. H. Freeman, 1969), pp. 62–105.

3. Van B. Boddie of Greenville, Mississippi, letter to the *Rural New Yorker*, September 9, 1909. Wickson papers, Library of the University of California, Davis.

4. H. W. Collingwood, letter to E. J. Wickson, September 14, 1909. Wickson papers, Library of the University of California, Davis.

5. Heiser, *Nightshades*, p. 105.

6. San Francisco *Chronicle*, August 10, 1909.

7. David Starr Jordan and Vernon Kellogg, *The Scientific Aspects of Luther Burbank's Work* (San Francisco: A. M. Robertson, 1909).

8. O. E. Binner, letter to E. J. Wickson, December 21, 1911. Wickson papers, Library of the University of California, Davis.

Fourteen: Pirates and Promoters

1. W. L. Howard, interview with Fred Suelberger, June 10, 1941. Howard papers, Bancroft Library, University of California, Berkeley.

2. *Luther Burbank: His Methods and Discoveries and Their Practical Application*, ed. Henry Smith Williams et al. (Santa Rosa: Luther Burbank Press, 1915), 12:254.

3. G. H. Shull, letter to W. L. Howard, November 25, 1939. Howard papers, Bancroft Library, University of California, Berkeley.

Fifteen: "I Love Everything!"

1. G. H. Shull, letter to W. L. Howard, November 25, 1939. Howard papers, Bancroft Library, University of California, Berkeley.

2. N. I. Vavilov, *The Origin, Variation, Immunity and Breeding of Cultivated Plants* (New York: Ronald Press, 1951).

3. T. D. Lysenko, *Heredity and Its Variability* (Moscow: Foreign Languages Publishing House, 1953), p. 94.

4. Paramahansa Yogananda, *Autobiography of a Yogi* (Los Angeles: Self-Realization Fellowship, 1973), p. 411.

5. Ibid., p. 415.

6. Burbank's lifelong temperance was a byword. In an interview with W. L. Howard in 1940, Santa Rosa banker Frank P. Doyle recalled an occasion when Burbank was prevailed upon to drink "a thimbleful or two" of liquor while taking time out from jury duty. When he and the other jurors returned to their rooms, "Luther proceeded to turn handsprings in his pajamas. His companions wondered what he would have done if he had had a real drink." Howard papers, Bancroft Library, University of California, Berkeley.

7. Ada Kyle Lynch, *Luther Burbank, Plant Lover and Citizen; with Musical Numbers* (San Francisco: Harr Wagner, 1924).

8. San Francisco *Examiner*, January 27, 1925.

9. Edgar Waite, "Luther Burbank, Infidel," *Haldeman-Julius Monthly* 3 (March 1926):490–99.

10. See L. Sprague de Camp, *The Great Monkey Trial* (Garden City, N.Y.: Doubleday, 1968). Burbank is supposed to have commented: "Mr. Bryan is an honored friend of mine, yet this need not prevent the observation that the skull with which nature endowed him visibly approaches the Neanderthal type. Feeling and the use of gesticulation and words are more according to the nature of this type than investigation and reflection" (p. 102).

11. Frederick W. Clampett, *Luther Burbank, "Our Beloved Infidel"* (New York: Macmillan, 1926), p. 36.

12. The quotation is from Olive Schreiner's *The Story of an African Farm* (London: Chapman & Hall, 1883).

13. Ben B. Lindsey, "My Address at the Grave of Luther Burbank," *Haldeman-Julius Monthly* 4 (June 1926):100–107. See also Maynard Shipley, "Luther Burbank's Last Rites" in the same issue, pp. 22–26.

14. C. H. Cramer, *Royal Bob: The Life of Robert G. Ingersoll* (Indianapolis: Bobbs-Merrill, 1952), p. 151.

Sixteen: Burbank's Legacy

1. Orland E. White, letter to W. L. Howard, December 21, 1938. Howard papers, Bancroft Library, University of California, Berkeley.

2. The Elberta peach was originated by Samuel H. Rumph of Marshallville, Georgia, around 1870. It was a seedling of a Chinese cling peach, the pollen parent being presumed to be an Early Crawford. This July Elberta was introduced by Stark Brothers after Burbank's death and was apparently a selected seedling from the Sebastopol orchard.

3. Thomas G. Sexton, national sales manager, Stark Brothers Nurseries and Orchards Company, letter to the author, October 16, 1974.

4. Hugo de Vries, "Burbank's Production of Horticultural Novelties," *Open Court* 20 (1906): 644.

5. Robert F. Scharpf and William McCartney, "*Viscum album* in California—Its Introduction, Establishment and Spread," *Plant Disease Reporter* 59 (March 1975): 258.

6. *The Eddy Tree Breeding Station: Institute of Forest Genetics.* Interviews conducted by Lois C. Stone, sponsored by the Forest History Society, Regional Oral History Office, University of California, Berkeley, 1974.

7. Dickson Terry, *The Stark Story* (St. Louis: Missouri Historical Society, 1966), pp. 84–87.

8. Edgar Anderson, *Plants, Man and Life* (Berkeley and Los Angeles: University of California Press, 1971), p. 81.

9. C. D. Darlington, "The Retreat from Science in Soviet Russia," in *Death of a Science in Russia*, ed. Conway Zirkle (Philadelphia: University of Pennsylvania Press, 1949), pp. 70–71.

10. H. J. Muller, "Genetics in Relation to Modern Science," in *Death of a Science in Russia*, p. 93.

11. Darlington, "Retreat from Science in Soviet Russia," p. 74.

12. Luther Burbank to J. H. Empson, letter dated February 29, 1908; J. H. Empson to George Shull, letters dated September 21 and 27, 1910, and April 16, 1914.

13. Donald F. Jones, "The Life and Work of Luther Burbank," unpublished manuscript, quoted by permission of the Connecticut Agricultural Experiment Station, pp. 51–52.

14. Walter L. Howard, "Luther Burbank: A Victim of Hero Worship," *Chronica Botanica* 9, no. 5/6 (Winter 1945–46): 476. Howard states: "A letter from the company, dated March 31, 1943, verifies the foregoing statements and adds that they have grown from 1,500 to 2,000 acres of the peas, annually, since 1908." There is some reason to think that this may not be entirely accurate, but the Burbank Admiral would appear to have been grown for at least a decade after its introduction.

15. Ralph B. Howe, letter to W. L. Howard, March 2, 1938. Howard papers, Bancroft Library, University of California, Berkeley.

16. F. Scott Fitzgerald, *The Crack-Up*, ed. Edmund Wilson (New York: New Directions, 1956), p. 69.

17. Ernst Mayr, *The Growth of Biological Thought: Diversity, Evolution, and Inheritance* (Cambridge, Mass.: Harvard University Press, Belknap Press, 1982), p. 824.

18. Stephen Jay Gould, "Triumph of a Naturalist," review of *A Feeling for the Organism: The Life and Work of Barbara McClintock* by Evelyn Fox Keller, *New York Review of Books*, March 29, 1984, p. 3

19. See, for example, Jack Fincher, "Tailored Genes," *Horticulture* 62 (April 1984): 50–57.

20. See Mayr, *Growth of Biological Thought*, p. 731.

21. Virginia Walbot and C. A. Cullis, "The Plasticity of the Plant Genome—Is It a Requirement for Success?" (Stanford University, Department of Biological Sciences, 1984, photocopy), pp. 2, 9–11. See also the following publications by C. A. Cullis: "Molecular Aspects of the Environmental Induction of Heritable Changes in Flax," *Heredity* 38 (1977): 129–54; "Quantitative Variation of Ribosomal RNA Genes in Flax Genotrophs," *Heredity* 42 (1979): 237–46; "Environmental Induction of Heritable Changes in Flax: Defined Environments Inducing Changes in rDNA and Peroxidase Isozyme Band Pattern," *Heredity* 47 (1981): 87–94; and "Environmentally Induced DNA Changes in Plants," *CRC Critical Reviews in Plant Sciences* 1, no. 2 (1984): 117–31. And see references in chapter 3, note 29, above, in connection with the "neo-Lamarckian tracking system" proposed by immunologist E. J. Steele.

INDEX

Designer: Laurie Anderson
Compositor: G & S Typesetters, Inc.
Text: 11/13 Sabon
Display: Sabon
Printer: Maple-Vail Book Mfg. Group
Binder: Maple-Vail Book Mfg. Group